Accounting, Finance, Sustainability, Governance & Fraud: Theory and Application

Series Editor

Kıymet Tunca Çalıyurt, Iktisadi ve Idari Bilimler Fakultes, Trakya University, Balkan Yerleskesi, Edirne, Türkiye

This Scopus indexed series acts as a forum for book publications on current research arising from debates about key topics that have emerged from global economic crises during the past several years. The importance of governance and the will to deal with corruption, fraud, and bad practice, are themes featured in volumes published in the series. These topics are not only of concern to businesses and their investors, but also to governments and supranational organizations, such as the United Nations and the European Union. Accounting, Finance, Sustainability, Governance & Fraud: Theory and Application takes on a distinctive perspective to explore crucial issues that currently have little or no coverage. Thus the series integrates both theoretical developments and practical experiences to feature themes that are topical, or are deemed to become topical within a short time. The series welcomes interdisciplinary research covering the topics of accounting, auditing, governance, and fraud.

Marcelo Jasmim Meiriño ·
Osvaldo Luiz Gonçalves Quelhas
Editors

Critical Factors Associated with Environmental, Social and Governance Management in Organizations

Springer

Editors
Marcelo Jasmim Meiriño
School of Engineering
Fluminense Federal University
Niterói, Rio de Janeiro, Brazil

Osvaldo Luiz Gonçalves Quelhas
School of Engineering
Fluminense Federal University
Niterói, Rio de Janeiro, Brazil

ISSN 2509-7873 ISSN 2509-7881 (electronic)
Accounting, Finance, Sustainability, Governance & Fraud: Theory and Application
ISBN 978-981-96-1724-1 ISBN 978-981-96-1725-8 (eBook)
https://doi.org/10.1007/978-981-96-1725-8

This Springer imprint is published by the registered company Springer Nature Singapore Pte Ltd.
The registered company address is: 152 Beach Road, #21-01/04 Gateway East, Singapore 189721, Singapore

Foreword

This book presents a contextualized and practical understanding of how Environmental, Social and Governance (ESG) can be used to improve organizational sustainability. Over the past decade, ESG has gained recognition for its ability to improve governance, image, credibility, and reputation, along with its competence to create new business opportunities, especially in the social and environmental areas. Based on these points, the book is relevant to the current context of sustainable governance.

The organizers, who are respected researchers with theoretical and practical knowledge of the subject, present contributions that provide an understanding of the topic from a practical application perspective. By facilitating the understanding, application, and use of ESG concept, this book adds value and becomes a valuable source of information, especially for practical purposes.

In this book, students and professionals in engineering, administration, and economics will find a comprehensive analysis of the critical factors for implementing ESG. Also, the contextualized approach to the topic makes the topic clear and understandable for those interested in ESG-based organizational management.

Therefore, I confidently recommend this book as one of the most relevant concerning the practical application of ESG concept.

Rio de Janeiro, Brazil

Cristiano Alexandre Virginio Cavalcante

Contents

Editors and Contributors

About the Editors

Marcelo Jasmim Meiriño is Professor at Fluminense Federal University (UFF) in School of Engineering; Professor of Doctoral Program in Sustainable Management Systems (PPSIG UFF); and Professor of Master Program on Management Systems (MSG UFF); Marcelo Jasmim Meiriño holds Ph.D. in civil engineering, emphasizing in management, production, quality, and sustainable development. Marcelo Jasmim Meiriño is Architect and Urban Planner; Master in Civil Engineering; Occupational Safety Engineer; Coordinator in the Innovation Center and Technology for Sustainability (NITS/UFF); Specialist in Sustainability and Energy Efficiency in Buildings; Member of the Social Responsibility Commission of the Brazilian Institute of Oil, Gas and Biofuels (IBP); Coordinator of the National Congress on Management Excellence (CNEG); Reviewer of journals and academic events; Researcher; and Consultant.

Osvaldo Luiz Gonçalves Quelhas is Coordinator of Technology, Business, and Environment Management Laboratory (LATEC UFF); Professor at Fluminense Federal University (UFF) in School of Engineering; Coordinator and Professor in Doctoral Program in Sustainable Management Systems (PPSIG UFF); Vice Coordinator and Professor in Master Program on Management Systems (MSG UFF); Ph.D. in Production Engineering, graduated as Civil Engineer; Member of the Social Responsibility Commission of the Brazilian Institute of Oil, Gas and Biofuels (IBP); Coordinator of the National Congress on Management Excellence (CNEG); he worked as Engineer and Process Manager in several national and international organizations such as the Brazilian Naval Arsenal in Rio de Janeiro, Mills Cover Letter Equipamentos Ltda and Shell Brasil S.A. Reviewer of journals and academic events; Researcher and Consultant.

Contributors

Bernardo Shogo Okada Ahmed Fluminense Federal University, Niterói, Brazil

Lidia Angulo-Meza Production Engineering, Federal Fluminense University, Niterói, Brasil

Marcio Silva Borges Technology and Innovation Sciences in Agribusiness, Program for Territorial Development and Public Policy (PPGDT), Federal Rural University of Rio de Janeiro (UFRRJ), Seropédica, Brazil

Pablo Carpejani Business of Pontifical Catholic, University of Parana (PUCPR), Curitiba, Brazil

Sergio E. Gouvea Da Costa Industrial and Systems Engineering of Federal University of Technology, Curitiba, Parana (UTFPR), Brazil

André Cantareli da Silva Civil Engineering, Department of Accounting Sciences, Federal Fluminense University, Niterói, Brasil

Sergio Ricardo Da Silveira Barros Fluminense Federal University, Niterói, RJ, Brazil

Nathália De Almeida Marques Fluminense Federal University, Niterói, Brazil; Fluminense Federal University, Niterói, Brazil

Flávio Ezequiel De Andrade Fluminense Federal University, Niterói, Brazil

Fernando Oliveira De Araujo Organizational Management, Production Engineering Department, School of Engineering, Fluminense Federal University, Niterói, Rio de Janeiro, Brazil

Pâmela De Carvalho Marques Silva Production Engineering, Federal Fluminense University, Niterói, Brasil

Edson Pinheiro De Lima Industrial and Systems Engineering of Federal University of Technology, Curitiba, Parana (UTFPR), Brazil

Juliana Damaris Candido De Lima Production Engineering, Federal Fluminense University, Niterói, Brasil

Eduardo De Oliveira Department of Production Engineering, Faculty of Engineering, Universidade Federal Fluminense, Niterói, Brazil

Raquel De Oliveira Moreira Management Systems, Fluminense Federal University, Niterói, Brazil

Letícia De Oliveira Gago Ramos De Souza Department of Production Engineering, Faculty of Engineering, Universidade Federal Fluminense, Niterói, Brazil

Micheli Santos De Souza Management Systems, Fluminense Federal University, Niterói, Brazil

Lincoln Campelo Dias Engineering Systems Management, Federal Fluminense University, Niterói, Brasil

Michelle Do Carmo Vieira Territorial Development and Public Policy, Program for Territorial Development and Public Policy (PPGDT), Federal Rural University of Rio de Janeiro (UFRRJ), Seropédica, Brazil

Felipe Rodrigues Do Nascimento Business Economics, Institute of Educational Development, Getulio Vargas Foundation, Rio de Janeiro, Brazil

Wagner Dos Santos Ramos Fluminense Federal University, Niterói, Rio de Janeiro, Brazil

Samuel Martins Drei Production and Manufacturig Engineering, Federal Fluminense University, Niterói, Brasil

Sergio Luiz Braga França Civil Engineering, Fluminense Federal University, Niterói, RJ, Brazil

Sergio E. Gouvea Da Costa Industrial and Systems Engineering, Federal University of Technology-Parana (UTFPR), Parana, Brazil

Ruben Huamanchumo Gutierrez Production Engineering, Federal Fluminense University, Niterói, Brasil

Rafael A. Kluska Fluxo Business Automation, Curitiba, Brazil

Gilson Brito Alves Lima Production Engineering, Federal Fluminense University, Niterói, Brasil

Fernanda Ribeiro Ramos Lopes Organizational Management, Management Systems Program, School of Engineering, Fluminense Federal University, Niterói, Rio de Janeiro, Brazil

Carla Gonçalves Machado Business Administration at Jönköping International Business School, Jönköping University, Jönköping, Sweden

Mirian Picinini Méxas Civil Engineering, Actuarial Sciences, Federal Fluminense University, Niterói, Brasil

Marcelo Jasmim Meiriño Doctoral Program in Sustainable Management Systems, Fluminense Federal University (UFF) in School of Engineering, Niterói, Brazil

Ligia Moreira Management Systems, Fluminense Federal University, Niterói, Brazil

Julio Vieira Neto Department of Actuarial Sciences and Finance, Fluminense Federal University, Niterói, Brazil

Renan Carriço Payer Production Engineering, Federal Fluminense University, Niterói, Brasil

Edson Pinheiro De Lima Industrial and Systems Engineering, Federal University of Technology-Parana (UTFPR), Parana, Brazil

Adriane Domingues Quelhas Fluminense Federal University, Niterói, Brazil

Osvaldo Luiz Gonçalves Quelhas Business and Environment Management Laboratory (LATEC UFF), Fluminense Federal University (UFF) in School of Engineering, Niterói, Brazil

Priscilla Cristina Cabral Ribeiro Production Engineering Department at Universidade Federal Fluminense (UFF), Rio de Janeiro, Brazil

Lizzie Bessa Risicato Production Engineering, Federal Fluminense University, Niterói, Brasil

Lucia Cristina Dantas Shpielman Management Systems, Fluminense Federal University, Niterói, Brazil

Luiz Perez Zotes Fluminense Federal University, Niterói, Brazil

List of Figures

List of Tables

Part I
Introduction

Chapter 1
ESG and Sustainable Development: A Blueprint for Ethical Investment and Organizational Growth

Osvaldo Luiz Gonçalves Quelhas and Marcelo Jasmim Meiriño

Abstract This content explores the principles of Environmental, Social, and Governance (ESG) as a framework for promoting sustainable development and ethical investment. With increasing global concerns such as the COVID-19 pandemic, environmental degradation, and socio-economic inequality, ESG has emerged as a vital tool for guiding organizations toward responsible practices. The concept of ESG integrates environmental stewardship, social responsibility, and sound governance, making it a key criterion in investment decisions. This study delves into the origins of ESG, its evolution, and its growing importance in corporate governance and the financial sector. Through the analysis of ESG metrics and their application, the paper underscores the role of ESG in shaping long-term value and organizational sustainability. Furthermore, it highlights the growing demand for transparency and accountability, noting how ESG indicators assess financial and non-financial performance. The findings suggest that organizations adopting ESG principles enhance their reputations and secure competitive advantages in today's investment landscape. By embedding ESG in decision-making processes, businesses can foster stronger relationships with stakeholders while contributing to the global effort of achieving sustainable development goals (SDGs).:

Keywords ESG—Sustainable · Development—Ethical · Investment—Organizational growth

O. L. G. Quelhas
Business and Environment Management Laboratory (LATEC UFF), Fluminense Federal University (UFF) in School of Engineering, Niterói, Brazil
e-mail: osvaldoquelhas@id.uff.br

M. J. Meiriño (✉)
Doctoral Program in Sustainable Management Systems, Fluminense Federal University (UFF) in School of Engineering, Niterói, Brazil
e-mail: marcelojm@id.uff.br

M. J. Meiriño and O. L. G. Quelhas (eds.), *Critical Factors Associated with Environmental, Social and Governance Management in Organizations*, Accounting, Finance, Sustainability, Governance & Fraud: Theory and Application, https://doi.org/10.1007/978-981-96-1725-8_1

1.1 Introduction

As established by the UN, knowledge regarding human rights and duties provides guidance for individuals' interactions with nature and society. It advises respect for nature and emphasizes the connection between people and the environment. The main idea is to use nature wisely and to seek a balance where both nature and human beings can thrive.

According to Li, Ting-Ting et al. (2021), the world faces significant challenges: the global spread of COVID-19, the collapse of the US stock market, the locust plague in Africa, and environmental (E), social (S), and governance (G) issues have raised global concern.

The theme of sustainable development has become a universal concern and has established itself as a focal point for discussions worldwide. In response to these challenges, international and national organizations have put forward action plans to promote sustainable development, such as ESG, to build a sustainable and universally applicable development framework.

The principles of ESG constitute a system with a structure that includes environmental (E), social (S), and governance (G) factors. ESG is defined as a form of responsible investment. The Principles for Responsible Investment (PRI) are described as: "a strategy and practice to incorporate environmental, social, and governance (ESG) factors into investment and ownership decisions" (PRI 2018).

ESG is a strategy investors use to analyze organizational behavior and monitor future financial performance. As an investment principle to analyze organizations' sustainable development, the three basic aspects of ESG are critical factors considered in investment analysis and decision-making.

Environmental, social, and governance (ESG) factors help measure the sustainability and social impact of organizational activities. The European Banking Authority (EBA 2021) defines ESG factors as "environmental, social, or governance issues that may have a positive or negative impact on the financial performance or solvency of an entity, whether sovereign or individual."

Therefore, ESG constitutes a dimension for analyzing the presence of sustainable development principles in projects and organizational operational strategies, taking into account economic, environmental, social, and governance outcomes. In summary, ESG is an investment philosophy that seeks long-term value growth and is a systematic, concrete, and realistic governance method.

According to Shen et al. (2023), business ethics and respect in relationships with stakeholders, non-discrimination, respect for diversity of opinions and people, and cooperation with suppliers and customers—considered organizational strategic objectives—define what is justifiable in the relationship between socio-environmental responsibility and economic outcomes.

There is an emphasis on encouraging government, private, and third-sector organizations to live in harmony with nature and society. The concept of ESG has emerged over the past two decades as an essential ethical framework. This concept encourages organizations to develop a balanced relationship with nature and society.

The ESG concept was first introduced in 2004 when Kofi Annan, the Secretary-General of the United Nations, proposed a discussion on how to promote ethical business practices and profit-making. The outcome, formulated by 18 financial institutions from nine countries in 2005, was a report titled "Who Cares Wins: Connecting Financial Markets to a Changing World." This report recommended that financial institutions consider ESG factors in their investment decisions.

In 2006, the United Nations established the "Principles for Responsible Investment" (PRI), which were designed to help investors understand the impact of ESG on investments and to emphasize the integration of ESG into investment practices within financial institutions.

Since ESG was formally proposed in 2004, it has been applied worldwide. Organizations and the UN have worked together to develop and mature the monitoring of environmental, social, and governance outcomes, both promised and realized. ESG has matured through the establishment of the ESG evaluation system, ESG disclosure standards, and the ESG index system.

As the ESG concept has become universally recognized, it has been widely studied, practiced, and popularized. Literature reviews on ESG have primarily focused on the "Governance" aspect, highlighting a shortage of studies on environmental (E) and social (S) factors. ESG encompasses an integrated structure of environmental, social, and governance factors and demonstrates an interactive relationship between its dimensions.

Many studies have focused on the interaction between environmental (E), social (S), and governance (G) factors. These factors play an important role in monitoring the future financial performance and social impact of organizations. Therefore, as an investment analysis principle that integrates environmental, social, and governance factors, ESG significantly contributes to driving sustainable development within organizations. The interactive relationship between its dimensions is also a critical aspect that deserves attention to promote the improvement of economic outcomes.

PRI defines ESG investment as a strategy that incorporates environmental, social, and governance factors into investment decisions, commonly referred to as "Sustainable Investment," "Ethical Investment," or "Impact Investment" (PRI 2018).

Countries' efforts to develop a sustainable economy still have a long way to go to achieve the Sustainable Development Goals (SDGs) proposed by the United Nations. The progress of various countries in ESG is a path to realize high-quality economic development.

The adoption of ESG principles by economic agents and organizations has stimulated academic research. Regarding ESG disclosure practices, sustainability reports are published as a resource or "tool" for measurement and disclosure to both internal and external stakeholders. These reports serve as a "tool" to monitor whether an organization meets the sustainable development goals.

In summary, environmental, social, and governance factors have gained increasing attention from organizations. This trend can be explained in two ways. First, it is related to historical and contextual compensation, where organizations engaged in a particular economic activity have negatively impacted the environment and society, consequently harming the community (people and environment) in which they are embedded. Second, it relates to value generation for the organization, as practices related to environmental, social, and governance factors are currently evaluated by the financial market to decide whether to invest.

Organizations with ESG indicators and strong environmental, social, and economic performance will be ahead of their competitors. Organizations are evaluated by their stakeholders based on both their economic performance and their performance concerning non-financial indicators (social and environmental indicators).

In this context, financial information is shifting toward an integrated approach, where environmental, social, and economic performance is part of assessing the quality and performance of the company. Reports derived from this information place ESG indicators as the focal point of the investment chain in financial markets.

1.2 ESG: Organizational Performance Indicator

According to Pereira et al. (2021), organizations tend to be evaluated by their stakeholders along two dimensions: (1) by their performance in the competitive market and (2) by their performance concerning non-financial values. ESG practices fall into the latter category. Analyzing ESG practices is useful in diagnosing the stage of maturity regarding ESG principles. ESG maturity involves the development of the organization in the realms of environmental, social, and governance responsibility.

On the other hand, the misuse of declared ESG information can lead to critical evaluations, harm transparency, and consequently undermine the credibility of the organization in terms of its ESG performance. Thus, the composition and declarations regarding the application of concepts related to socio-environmental responsibility and good governance practices should be structured to be monitored and identified according to transparency criteria.

Drempetic et al. (2019) present evidence of positive correlations between organizational size, accessibility, and the availability of data that influence ESG metrics. There are numerous ways to measure an organization's performance, especially given the diversity of data, and it is essential that each ESG factor includes financial indicators. These indicators are directly related to economic performance and are understood as the quantitative translation of corporate strategy (Kocmanová et al. 2013).

Non-financial indicators should be identifiable according to the organization's objectives and support the monitoring of ESG performance. Kocmanova and Simberova (2012) emphasize that these qualitative ESG performance indicators should be useful in supporting decision-making regarding investments.

Some examples of ESG indicators include:

1. **Environmental factor:** energy consumption, use of potable water, resilience to climate change, environmental policy, land use, natural resource management, hazardous waste and material management, among others.

2. **Social factor:** consumer rights, corporate philanthropy, data security and customer privacy, diversity issues, employee engagement, community health and safety, human rights, responsible people management, human capital management.
3. **Governance factor:** accounting and consulting, board composition, business ethics, compliance, executive compensation, corporate structure, transparency, fund governance, political contributions, reporting and disclosure, succession planning.

Economic performance indicators provide quantitative monitoring and reflect the results of corporate strategies (Kocmanová and Dočekalová 2012). Non-financial metrics and KPIs reflect the nature of the organization, offering a broader view of economic performance.

1.3 ESG as an Investment Decision-Making Tool

The search for indicators related to the environment, social well-being, and governance is being driven forward, structuring logical and rational approaches for decision-making. Investors aim for financial benefits but within an ethical and responsible investment environment.

The application and identification of ESG principles in organizational practices reveal a corporate governance structure composed of good practices, which consequently lead to strong overall performance (Bektur and Arzova 2020). When represented by non-financial indicators, ESG requires consistent analysis by asset managers. Organizations are accounting for both financial and non-financial performance as they are motivated to become responsible and transparent for investors and other stakeholders (Camilleri 2017).

The adoption of ESG indicators and accountability to stakeholders reduces organizational risk and capital cost. According to Kocmanová and Dočekalová (2012), the emphasis on accountability will gradually shift from a financial indicator to an integrated approach, where information on environmental, social, and economic performance will be part of the assessment of an organization's quality and performance.

ESG information is associated with various economically significant outcomes, according to Amel-Zadeh and Serafeim (2018). The availability of information enables appropriate and accurate monitoring of sustainability management performance.

1.4 ESG as a Socially and Environmentally Responsible Investment

Many organizations currently measure and report environmental data related to their production (e.g., carbon emissions, water consumption, waste generation), social data (e.g., employee information, product information, customer-related information), and governance data (e.g., political lobbying, anti-corruption programs, board diversity).

According to Prorokowski (2016), organizations begin implementing ESG indicators associated with operations, with technical care and adherence to logic aimed at making sound decisions. The demand for evaluating ESG indicators and certifying environmental, social, and governance indicators can weaken an organization's positioning by making it subject to external approval.

The need to address climate change has emerged in the evaluation practices of ESG indicators. It is a natural consequence and product of the progress of the ESG approach, which integrates factors and establishes a scenario that enables the analysis of complex problems. Organizations value reporting their environmental performance, aware that it enhances their reputation with stakeholders.

One of the reasons organizations publish environmental reports is to signal to stakeholders that they have adequate environmental performance and to demonstrate that they value and prioritize sustainable development, according to Jaafar et al. (2019).

1.5 ESG as Practice in Organizations

According to Giamporcaro and Leslie (2018), the practice and awareness related to ESG support a set of data to be presented to stakeholders and financial analysts. In general, the data are analyzed and used to verify the credibility of organizational sustainability.

Organizations develop their own methodologies and sets of indicators. The result is a complex ecosystem of ESG metrics, with ratings that need to be contextualized for correct interpretation. To enable accurate analyses and decisions, professionals analyzing ESG management reports need to have a comprehensive understanding of these compositions and indicator metrics.

Beretta et al. (2021) highlight the current trend in reports suggesting that organizational competitiveness is directly linked to the ability to establish financial and non-financial indicator metrics accurately, correctly, and logically. Baraibar-Diez and Odriozola (2019) confirm that there is significantly greater non-financial performance in organizations that develop a Corporate Social Responsibility (CSR) Committee.

One way to mitigate interpretation risks related to social and environmental indicators associated with financial indicators is to find ways to include stakeholders by creating committees or structures that involve them and ensuring they participate actively.

For good ESG management operations, intelligent management of the risks and opportunities arising from its implementation is essential.

References

Amel-Zadeh A, Serafeim G (2018) Why and how investors use ESG information: evidence from a global survey. Financ Anal J 74(3):87–103

Baraibar-Diez E, D Odriozola M (2019) CSR committees and their effect on ESG performance in UK, France, Germany, and Spain. Sustainability 11(18):5077

Bektur Ç, Arzova S (2020) The effect of women managers in the board of directors of companies on the integrated reporting: example of Istanbul stock exchange (ISE) sustainability index. J Sustainable Finance and Invest

Beretta V, Demartini MC, Lico L, Trucoo S (2021) A tone analysis of the non-financial disclosure in the automotive industry. Sustainability 13:2132

Camilleri MA (2017) The integrated reporting of financial, social and sustainability capitals: a critical review and appraisal. Int J Sustain Soc 9(4):311–326

Drempetic S, Klein C, Zwergel B (2019) The influence of firm size on the ESG score: corporate sustainability ratings under review. J Bus Eth 1–28

EBA (2021) EBA report on management and supervision of ESG risks for credit institutions and investment firms. Available online: https://www.eba.europa.eu/sites/default/documents/fifiles/document_library/Publications/Reports/2021/1015656/EBA%20Report%20on%20ESG%20risks%20management%20and%20supervision.pdf (Accessed on 31 July 2024)

Giamporcaro S, Leslie D (2018) Responsible investment at old mutual: a case of institutional entrepreneurship. Emerald Publishing Limited. V 8(4):1–29

Jaaffar AH, Alrazi B, Ooi S, Shamsuddin A (2019) Strategically-framed environmental disclosure index: a measurement approach of Malaysian public listed companies' corporate environmental reporting practices. Int J Environ Technol Manage 22(4–5):236–256

Kocmanova A, Simberova I (2012) Modelling of corporate governance performance indicators. Eng Econom 23(5):485–495

Kocmanová A et al (2013) Construction of the economic indicators of performance in relation to environmental, social and corporate governance (ESG) factors. Acta Universitatis Agriculturae Et Silviculturae Mendelianae Brunensis 60(4):195–206

Kocmanova A, Dočekalová M (2012) Construction of the Economic Indicators of Performance in Relation to Environmental, Social And Corporate Governance (Esg) Factors. Acta Universitatis Agriculturae et Silviculturae Mendelianae Brunensis volume lX 21 Number 4, 2012.

LI T-T et al (2021) ESG: research progress and future prospects. Sustainability 13(21):11663

Pereira RI, Marcilio B, Guercio M, Takimoto T, Fialho F (2021) ESG: Uma revisão integrativa. ENGEMA 2021

Principles of Responsible Investment (PRI) (2018) https://www.unpri.org/annual-report-2018 (Accessed on 1 March 2025)

Prorokowski L (2016) Environmental risk index for financial services firms. Qual Res Fin Markets

Shen H et al (023) ESG in China: a review of practice and research, and future research avenues. China J Account Res 100325

Osvaldo Luiz Gonçalves Quelhas is Coordinator of Technology, Business, and Environment Management Laboratory (LATEC UFF), Professor at Fluminense Federal University (UFF) in

School of Engineering; Coordinator and Professor in Doctoral Program in Sustainable Management Systems (PPSIG UFF); Vice Coordinator and Professor in Master Program on Management Systems (MSG UFF); Ph.D in Production Engineering, graduated as Civil Engineer; Member of the Social Responsibility Commission of the Brazilian Institute of Oil, Gas, and Biofuels (IBP); Coordinator of the National Congress on Management Excellence (CNEG). He worked as Engineer and Process Manager in several national and international organizations such as the Brazilian Naval Arsenal in Rio de Janeiro, Mills Cover Letter Equipamentos Ltda and Shell Brasil S.A. Reviewer of journals and academic events; Researcher and Consultant.

Marcelo Jasmim Meiriño is Professor at Fluminense Federal University (UFF) in School of Engineering; Professor of Doctoral Program in Sustainable Management Systems (PPSIG UFF); Professor of Master Program on Management Systems (MSG UFF); Ph. D in Civil Engineering, emphasis in Management, Production, Quality and Sustainable Development; Architect and Urban Planner; master's in civil engineering; Occupational Safety Engineer; Coordinator in the Innovation Center and Technology for Sustainability (NITS/UFF); Specialist in Sustainability and Energy Efficiency in Buildings; Member of the Social Responsibility Commission of the Brazilian Institute of Oil, Gas and Biofuels (IBP); Coordinator of the National Congress on Management Excellence (CNEG); Reviewer of journals and academic events; Researcher and Consultant.

Part II
Environmental Perspective

Chapter 2
Fundamentals of the ESG Concept from the World of Finance to Sustainable Business Management

Julio Vieira Neto, Luiz Perez Zotes, Flávio Ezequiel De Andrade, Adriane Domingues Quelhas, Nathália De Almeida Marques, and Bernardo Shogo Okada Ahmed

Abstract Faced with climate issues, governments, and shareholders face challenges imposed by intangible risks that are difficult to measure regarding economic aspects. In this sense, corporate reports need to meet the demands imposed by the new way of connecting with the theme of sustainability aligned with social, environmental and governance (ESG) issues. The ESG theme was introduced in the publication Who Cares Wins (The Global Compact 2004) by the Global Compact in partnership with the World Bank. The publication brought recommendations from the financial sector to integrate better environmental, social and governance issues in analysis, asset management and securities brokerage, as there is evidence that highlights the influence and importance of ESG management in the company's business, especially concerning value generation. In this sense, this chapter aims to present the topic debated in the ESG line reflected in a line of thoughts of the actors mentioned concerning the case studies presented. Given the debates presented, the chapter presents the importance of strategically analyzing the topic of sustainability in the face of intangible risks that affect the results of companies and governments that are difficult to measure.

Keywords Environmental · Social · Governance · Financial · Investments · Corporate disclosure

---

J. V. Neto (✉)
Department of Actuarial Sciences and Finance, Fluminense Federal University, Niterói, Brazil
e-mail: julion@id.uff.br

L. P. Zotes · F. E. De Andrade · A. D. Quelhas · N. De Almeida Marques · B. S. O. Ahmed
Fluminense Federal University, Niterói, Brazil
e-mail: flavioezequiel@id.uff.br

B. S. O. Ahmed
e-mail: bernardoshogo@id.uff.br

M. J. Meiriño and O. L. G. Quelhas (eds.), *Critical Factors Associated with Environmental, Social and Governance Management in Organizations*, Accounting, Finance, Sustainability, Governance & Fraud: Theory and Application, https://doi.org/10.1007/978-981-96-1725-8_2

2.1 Introduction

2.2 Role of the Chapter in the Book Structure

This chapter contributes to the thematic structure of the book in the face of ESG aspects in business and financial management from the perspective of creating value in organizations.

The environmental, social, and governance criteria, represented by the acronym ESG, emerged in 2004 from a provocation by the UN Secretary-General, Kofi Annan. He challenged 50 CEOs of large financial institutions on how to integrate social, environmental, and governance factors in the capital market, in a publication of the Global Compact in partnership with the World Bank, called Who Cares Wins.

In this sense, the sustainability agendas of several non-governmental bodies (United Nations Sustainable Development Goals, for example) began to bring pressure on companies and governments to take a stand. Consequently, each continent began to take an appropriate stance with the urgency of adapting to a new reality of future scarcity and risks that have never been faced before.

Therefore, the concept of sustainability in the business environment has never become so necessary. After all, "sustainability is the search for a balance between the supply of human needs and the preservation of natural resources, without compromising future generations" (Brundtland Report 1987).

Additionally, it is particularly important that with the advent of the Covid-19 pandemic, the importance of ESG, in terms of the social aspect, for the financial market has been put into perspective.

2.3 Historic of the Development of the Theme and Contemporary Issues

ESG has been gaining ground in the strategic field of organizations from different sectors due to the importance given by shareholders who finance companies. Therefore, the ESG orientation meets the search for less environmental and social impact, which has affected several organizations, generating financial liabilities that impact investors through lower dividends and market value.

Given this perspective, the ESG theme has been configured as an important front of analysis in the mitigation of risks and selection of companies with long-lived and resilient models to imminent changes, according to a study carried out by Whelan et al. (2020), in which it consolidates more than a thousand articles relating the correlation between good ESG practices and companies' financial performance. A negative correlation occurs in only 8% of the analyzed sample. Different from the traditional methods of evaluating companies such as DCF, Multiples, and Real Options, ESG is characterized as a complementary study front and intrinsic to the market's evolutionary process.

Damodaran (2021) in his article "Doing Good or Sounding Good A Skeptical Look at ESG" argues that ESG is a new and revolutionary idea that, in reality, does not have a major impact on the valuation (financial valuation) of companies. In his interpretation, it is more of a marketing and capital raising strategy for large financial institutions, as it is a topic of the moment. This approach will be discussed later in this work.

Faced with these issues, ignoring actions aimed at ESG criteria can pose risks to the environment, and legal or reputational risks, which can harm the organization and affect its financial results. On the other hand, companies with a strong ESG orientation tend to have a more stable and loyal investors base, thus obtaining better access to financing in the face of lower risk.

ESG aspect has been gaining ground in the strategic field of organizations from different sectors due to the importance given by shareholders who finance companies. Given this aspect, it is clear that the ESG orientation is in line with the search for less environmental and social impact, which have been affecting several organizations, generating financial liabilities that impact investors through lower dividends and market value.

Considering these aspects, ESG criteria are essential for companies that seek continuity, growth, and relevance. Society today has a more holistic view of organizations, in which the life cycle of its products, the relationship with the community, adoption of practices of diversity, gender equity, employee well-being, and environmental impacts, among others, are now valued.

Regarding this subject, studies have been carried out in countries such as the USA and the UK, followed by Europe (with a focus on Germany) and Japan; and in developing countries such as China, Korea, India, Eastern European countries such as Poland, and some African countries. In general, they point out that the adoption of good corporate governance practices positively affects corporate performance and/or company value, thus reducing the cost of capital. Consequently, companies with more robust governance have a lower cost of capital than those with weaker governance (Mishra and Kapil 2016).

Because of the environmental and social aspects, it is worth emphasizing the importance of these aspects as intangible assets that provide value to the organization. Therefore, the aspect of social and environmental capital emerges as value creation assets in the organization.

García-Merino et al. (2014) claim that the importance attributed to intangible resources has generated increasing interest, which encourages researchers to focus on knowledge and its management.

In this sense, intangibles have become the most basic resource to generate a competitive advantage. For Bueno et al. (2014), knowledge and management of intellectual capital have become key assets for companies to build sustainable competitive advantage. The scenario rapidly changes the knowledge, and therefore, the management of the company's intellectual capital. These are critical factors in taking advantage of opportunities.

Regarding the aspects linked to the creation of value in the face of intangible assets, Blair and Wallman (2001) add that they are non-physical factors that contribute to or are used in the manufacture of products or provision of services, or that generate the expectation of future profits to the companies or individuals who control its use.

In 2002, the definitions of Gu and Lev (2001) were reaffirmed, even in guidelines, as in the attribution given by MERITUM (2002) to intangible assets: they are non-monetary sources of probable future profits, which do not have a physical structure, are controlled (or at least influenced) by the results of a company's transactions and may or may not be sold separately from other assets.

Given the concepts of intangible assets, it is worth mentioning that the acronym ESG is directly related to the intangible aspects of the organization regarding the social as an important premise of this new tripod. In this sense, Roos and Roos (1997) state that social capital is the necessary factor for the success of an organization. Nahapiet and Ghoshal (1998) also add that the most valuable knowledge is in social interactions and relationships. There are many definitions for this capital that have been so fundamental for companies in recent decades according to the authors.

In Petzick et al. (1999) the authors, define social capital as the capacity for socialization that emerges from the relationship of trust and common norms to reduce costs with economic transactions and accelerate the pace of co-prosperity among cooperative partners.

Furthermore, social capital can be interpreted, according to Kang and Snell (2009) as the patterns of internal and external relationships between employees and other agents, acting as a relevant mechanism for the transfer and exchange of knowledge. This last definition is in line with the thinking of Nonaka and Takehuchi (1995), who state that social capital increases a company's ability to explore and generate intellectual assets. The ability to share knowledge, both explicit and tacit, is conditioned by the social structures that exist in an organization.

A particularity of this intangible asset identified by Burt (1992) is that, unlike other forms of capital, social capital is jointly owned by the parties in a relationship and neither party has or is capable of having exclusive property rights. This asset still plays a crucial role in creating both incremental and disruptive innovations and is closely linked to human capital (Fernández-Pérez et al. 2017). Therefore, the first step in creating a relevant social asset is to build sufficient interaction skills for each individual (Subramaniam and Youndt 2005).

The first example is related to the impact of social capital on small businesses that are part of cooperatives that have the advantages of being part of small groups, where interpersonal relationships provide self-monitoring mechanisms to keep them focused on goals that will benefit the whole group (Olson 1971). This organizational form, therefore, contains high levels of social capital that foster trust and supportive social networks (Meador et al. 2016).

Cook et al. (2013) state that the performance of cooperatives, using the intangible asset social capital, allows measuring the level of communication between members and the board. In addition to communication channels, voting policies and meeting format were identified as essential requirements in the ability of boards to obtain feedback from members, which, in turn, is considered critical to their performance.

The second evidence of the applicability of social capital was studied by Pechlaner and Pechlaner and Bachinger (2014) in the regional context. They verify that the business advantages generated by social capital in a regional context exist in a sustainable perspective and conclude that the company's regional integration is related to its competitive advantage.

For the authors of the literature, social capital can be defined as valuable, rare, difficult to imitate, and usable in different contexts and therefore meets the criteria for defining a core competence. As core competencies were defined as the basis for sustainable competitive advantage, social capital has the power to foster regional development, expressed through sustainable business activities. Social capital can, therefore, be understood as a promoter of entrepreneurship, maintaining a competitive region for long periods.

Another issue related to ESG is environmental issues, where companies that adopt environmental, social, and governance policies in their strategy contribute to building a more sustainable world and gain visibility in the capital markets, both by building a better corporate governance reputation and the achievement of better financial results. This is evident when comparing the return, volatility, and risk exposure of the companies that make up the ISE, with those that make up the IBOVESPA index.

However, the practice of environmental transparency can result in the organization being ranked low by institutions specialized in alerting the market about companies that are committed to good environmental practices.

However, companies poorly evaluated concerning environmental, social, and governance issues address actions that direct the organization toward sustainability. For example, Russian oil and gas companies, poorly evaluated, improved their performance with actions by their executives about the material benefits of adjusting to the ESG rating, presenting materiality consistent with actions in the social and environmental field (Clementino and Perkins Richard 2021).

One of the environmental impacts of the oil activity is the emission of greenhouse gases (GHG), present in the entire life cycle of the product, that is, from the well to the wheel. Although the tank-wheel phase has the highest emissions, oil tankers also contribute significantly to GHG emissions from the logistics sector. In this sense, it is up to this sector to seek technological alternatives to reduce its emissions through the decolonization of its production chain.

However, projects aimed at combating environmental impact have a greater focus on safety in the oil sector, seeking, for example, to prevent leaks. However, some solutions consist of changing maritime routes that can reduce the impacts of greenhouse gas emissions in the oil industry, and minimize the effects of the equivalent Co_2 (Greene et al. 2020).

2.4 CO_2 and ICT Emission Cases

How Do Companies Respond to Environmental, Social and Governance (ESG) ratings Evidence from Italy-annotated—Itália—2020
ESG ratings are "assessments" of a company based on a comparative assessment of its quality, standard, or performance on environmental, social, or governance issues. Corporate responses depend on managers' beliefs regarding the material benefits of fitting and scoring well on ESG ratings and their alignment with corporate strategy. There is strong evidence that US companies that received low scores on environmental assessments improved their environmental performance more than peers that were never assessed or that received a more positive assessment Likewise, Shvarts et al. (2018) show that oil and gas companies operating in Russia disclosed more information related to the environment after the introduction of a third-party environmental classification, with some adjustments to the format in which they reported the data

---

Does an Asset Owner's Institutional Setting Influence Its Decision to Sign the Principles for Responsible-annotated
Regarding Responsible Investments, in 2006, with the support of the United Nations, the Principles for Responsible Investment were launched, to unite investors in the search for responsible investment (IR). In 2016 there were more than 1,500 paying signatories and today (2021) there are more than 3,000
Why would Asset Owners be the motivators of the Responsible Investment market? - The universal owner hypothesis proposes that the largest institutional investors are investing in the market as a whole and therefore their portfolio will benefit more from a stable and growing global economy than from one stock outperforming another, taking advantage of externalities, such as environmental damage or destructive impact on communities (Hawley and Williams 2000, 2007; Monks 2001)
In a 2019 article, it was found that normative and cultural-cognitive institutional aspects, and to a lesser extent voluntary regulation, positively influence institutional investors in adopting the Principles for Responsible Investment (PRI). On the other hand, the practice of mandatory regulation reduced the probability of institutional investors adopting the Principles for Responsible Investment (PRI). Findings on normative and cultural-cognitive aspects build our understanding of how cultural factors, values, and beliefs are key drivers of CSR and RI adoption in general and in the context of the asset owner in particular. The symbolic meaning of PRI adoption makes it a natural reaction to expectations around investor behavior (normative institutional aspect), as well as a natural consequence of an alignment with the widely accepted values of the society in which the investor operates (the cognitive-cultural aspect). Pension funds that serve societies that exhibit more collectivist and humane attitudes are therefore more likely to sign up for a voluntary collaborative initiative with the stated aim of dealing with social issues and externalities that destroy our collective environmental heritage (Bernhagen and Mitchell 2010)

Influence of ESGC Indicators on Financial Performance of Listed Pharmaceutical Companies-annotated—Espanha—2021

Pharmaceuticals, although essential for human health, have a significant negative effect on the environment. Pharmaceutical production is associated with high consumption of raw materials, solvents, and energy, producing a high volume of waste and, in many cases, heavily polluting, affecting health through water pollution. Furthermore, once consumed and excreted, pharmaceuticals can enter the environment and be found in groundwater and soils, with harmful effects on aquatic organisms. That is why the pharmaceutical industry has increased its concern for the environment and, consequently, has adopted sustainable practices

This study analyzes the relationship between environmental, social, indicators, and financial performance, measured through ROA (Return On Assets), ROE (Return On Equity), and Tobin's Q, which are applied companies listed on the Nasdaq Smart Pharmaceuticals Index

Using structural equation modeling, the results confirmed the positive relationship between the construct composed of environmental, social, and governance (ESG) indicators and the aforementioned financial indices, concluding that investing in ESG is a profitable strategy

---

Would Overconfident CEOs Engage More in Environment, Social, and Governance Investments with a Focus on Female Representation on Boards -annotated- Coreia do Sul—2021

In recent studies, companies that issue their initial sustainability reports are reported to benefit from a substantial reduction in the cost of equity and better performance in ESG investments (Dhaliwal, D.S.; Li, O.Z.; Tsang, A.; Yang, Y.G). Studies show that companies that care about environmental performance are associated with lower costs of equity (Ghoul et al. 2011)

Studies also show that bank borrowing costs for companies with better corporate social responsibility (CSR) performance are lower than for more irresponsible companies, suggesting that encouraging ESG investments will lead to lower debt capital costs (Goss and Roberts 2011)

Finally, value investors are increasingly convinced that the more ESG investments focus on corporate sustainability and green growth, the more they will eventually reap the long-term benefits. When a company invests in ESG, it can be said that it is an act of paying attention to the external environment that surrounds it and signaling a willingness to try to be good for others. This is because ESG investments are incurring a high cost in the short term, so managers may be reluctant to invest

However, companies investing in ESG will be able to reduce the long-term financing costs that may arise from perceived default risk, signaling the company's willingness to pursue sustainability, even at the expense of the immediate cost. This signaling impact will be more pronounced for companies with opaque information environments that are subject to more information asymmetry. In conclusion, previous studies emphasize that ESG investments reinforce corporate intangible assets such as reputational capital, which leads to long-term investor wealth creation

Applying Responsible Ownership to Advance SDGs and the ESG Framework, Resulting in the Issuance of Green Bond-annotated—Islândia/ Canadá—2021

We apply a unique case research project from Reykjavik Energy (Orku veita Reykjavíkur, Reykjavík, Iceland), an Icelandic energy and utilities company, and use it to specifically analyze the links between ownership strategy, ODS, ESG, and bond issuance. green. After near bankruptcy in the years following the 2008 financial crisis, Reykjavik Energy formulated an ownership strategy that has become an important part of the company's resuscitation, along with a comprehensive financial rescue plan. Later, the company set relevant targets in line with the UN SDGs as well as an ESG methodology

In 2019, the company issued green bonds for the first time, which reflected the company's mission to put respect for the environment on an equal footing with profits. Green bonds received the highest possible rating—a "dark green shade"—from the independent body Center for International Climate Research, Oslo, Norway (CICERO). Reykjavik Energy followed up with a second green bond in 2020, and heading into 2021, the company's situation was markedly different from that in the years before and after the 2008 financial crisis; is considered an international example for simultaneously providing financial value to its three owners (three separate municipal governments) and essential services to constituents, while striving for complete environmental sustainability

The guiding vision of the ownership strategy summarizes the company's overall objective: "respect the environment, responsible use of natural resources and responsible use of capital" (Hale and Roger 2014). By using the concepts of Casadesus-Masanell and Ricart (2020) of "choices" and "consequences" in business models, we demonstrate how ownership strategy has become an essential reinforcement of a virtuous cycle, creating linkages to focus on domestic energy and geothermal infrastructure, responsible use of capital and maintenance of sustainable development criteria through SDGs and an ESG framework. All three options help the company fulfill its commitment to "respect the environment" while ensuring low-cost access to electricity and water. However, we started to see the virtuous cycle come to life with the introduction of the green bond

The green bond offers a less expensive financing option while ensuring that the capital is invested in responsible projects. Cheaper financing allows the company to deliver a stronger return on capital through the payment of dividends to municipal owners while allowing it to meet one of its key objectives in the ownership strategy: "Ensure services to its customers." Customers at reasonable and competitive prices. Since the green bond forces a high level of reporting on the application of funds to environmental resources, two-way enforcement takes place between the SDGs/ ESG and the green bond. As such, the green bond provides an important reinforcement between the SDGs/ESG and the responsible use of capital

Well-to-tank carbon emissions from crude oil maritime transportation—EUA/Noruega/ Espanha—2020 (pegada de carbono)

The oil and gas sector is responsible for significant greenhouse gas (GHG) emissions. Emissions arise both from the extraction, processing, transportation, and distribution of fuel (often referred to as the well-to-tank phase of the fuel lifecycle), as well as the eventual combustion of the fuel in various applications such as power, heat, and transportation. (El-Houjeiri et al. 2013; Greene and Lewis 2019)

Together, these stages form the well-to-wheel fuel lifecycle. While tank-wheel emissions are the main climate impact of oil and gas, the other elements of the fuel life cycle are important to account for and monitor (Rahman et al. 2015; Di Lullo et al. 2016)

The article focuses on carbon dioxide (CO_2) emissions, one of the most important elements of the well-to-tank stage: the international maritime transport of crude oil. The main means of transport for the intercontinental movement of oil is the oil tanker, largely powered by marine diesel and heavy fuel oil (Jia 2018)

The maritime sector is an important element of the low carbon transition strategy and was identified by the Energy Transitions Commission (ETC) as one of the most difficult sectors to decarbonize (ETC, 2018). The International Transport Forum (ITF) estimated that maritime transport was responsible for 3% of global emissions and 27% of emissions from freight transport in 2015, totaling around 873 million tons of CO_2 per year (ITF 2019)

Oil tankers account for 13% of maritime emissions, or approximately 114 million tons of CO_2 (Olmer et al. 2017). While reducing carbon emissions has become a focus for the shipping industry and its many customers, sustainability efforts in oil transportation have focused on safety issues—that is, preventing oil spills (Poulsen et al. 2016; Smith et al. 2012)

The importance of well-to-tank emissions is becoming more prominent with the increasing adoption of alternative fuels such as biodiesel or hydrogen (Bouman et al. 2017; Di Lullo et al. 2016; Ozawa et al. 2017; Winebrake et al. 2007), where emissions are often found mainly in the well-tank phase. Companies seeking to understand the true impacts of their activities and align with carbon accounting mandates and Science-Based Targets initiatives should include values for well-to-tank emissions (Greene and Lewis 2019; SBT 2018). Companies do this by using standard emission factors to convert fuel use into greenhouse gas emissions (Greene and Lewis 2019). Typically provided by government agencies or academic studies, emission factors are usually presented as a static value for well-to-tank and tank-to-wheel emissions, or combined as well-to-wheel emissions, for diesel, gasoline, and other fuels (Edwards et al. 2014; DEFRA 2019; EPA 2014). These factors are rarely provided based on fuel source or place of use. Proper accounting for carbon emissions is becoming increasingly important for companies, governments, and international government bodies as part of efforts to keep global temperatures below 2 degrees Celsius since pre-industrial times. These efforts need multi-stakeholder participation to join forces to achieve environmental improvements (see, for example, Schleifer 2013; Hale and Roger 2014; Abbott et al. 2015; Graham and Thompson 2015; Raza 2020)

As a result, companies in many industries are increasingly calculating, reporting, and seeking to reduce carbon emissions along their value chains, including the production and distribution of the fuels they consume. This is reflected in the growing trend toward including well-to-tank emissions in carbon accounting standards and climate target setting

In this study, we demonstrate the difference in tanker efficiency along major trade routes based on micro-level oil shipping data. We also demonstrate how this affects the carbon footprint of the oil cargo carried by these tankers, adding new dimensions to the work done by El-Houjeiri et al. (2013) and Masnadi et al. (2018) who consider oil life cycles using global industry average emission intensity values for oil tankers

The results show that the main emission factor is distance, despite optimized loads and more energy-efficient oil tankers. This suggests that efforts to reduce these emissions should be directed first at increasing local shipments rather than improving the Existing Vessel Design Index or loading factors. New shipping routes can lead to a reduction in emissions from oil transport, decreasing the distance that shipments need to travel, depending on naval and commercial viability (ITF 2019). Most significantly, the Kra Canal through the Malay Peninsula would shorten the range of trade between the Persian Gulf and East Asia by 1200 km. The Nicaragua Channel and the newly ice-free Arctic shipping lanes can also

There is potential to use the CO2e/ liter emission factors from this study for maritime oil transport to refine estimates of well-to-tank emissions on a commercial basis to more closely adjust to emissions from oil consumed by companies. This work would contribute to a growing need to understand well-to-tank emissions as alternative transport fuels with emissions primarily at the well-to-tank phase begin to be used more widely

In addition, the CO_2e/ton-km values are based on work by Clean Cargo, which offers commercial range factors for container shipping, creating a similar commercial range dataset for oil tankers. There is also potential to leverage this information to inform fueling decisions by governments or companies based on results, or as a factual basis to influence oil shippers to reduce emissions from their ships

Additional work could be done to characterize the other transport emissions that are also part of well-to-tank emissions, such as trucking, pipelines, or other transport activities that invariably occur as crude oil is further processed and distributed. As companies and governments look at their net-zero targets and the Paris Agreement, it is clear that including transport in the fuel lifecycle based on origin and destination is an important consideration that can help refine estimates of emissions and inform acquisition strategies

Emerging Corporate Disclosure of Environmental Social and Governance (ESG) Risks An Australian Study-annotated—Austrália—2020
In today's globalized business environment, companies around the world are under pressure to be accountable and transparent to society about their activities that can have a significant impact on the environment and society. In addition to what is disclosed in financial reports, the disclosure of ESG information has attracted the attention of corporate investors, especially institutional investors who tend to look to longer investment horizons (Lokuwaduge and Heenatigala 2017)
Indicators that reflect on the impact of ESG issues are essential for an analysis of a company's ability to maintain a long-term competitive advantage. This creates the need for companies to disclose information about their sustainability-related activities (Soderstrom 2013)
Sustainability reports are an integral part of communication between the company and its main stakeholders, including investors (Sjöström and Welford 2009). Soderstrom (2013) refers to sustainability reporting as the communication tool that companies use to publicize their corporate social responsibility (CSR) concerns and activities, including social and environmental impacts, in addition to the company's financial performance
Sustainability reporting is a process that helps organizations understand the links between organizational performance measurement and the organization's plans and strategy while making changes to address sustainability-related opportunities and threats (Global Reporting Initiative 2013). This process combines a company's profitability with corporate social responsibility towards its stakeholders. It is a more forward-looking business approach that creates long-term shareholder value by embracing the opportunities and managing the risks derived from economic, environmental, and social development

2.5 Summary of Main Ideas and Concepts

ESG issues remain a relevant topic in the main discussions on business sustainability. Therefore, the relationship between good corporate sustainability practices in the ESG pillars and their correlation with the financial performance of companies is being debated by organizations with a new frontier of sustainability

The environmental, social, and governance criteria, represented by the acronym ESG, emerged in 2004 from a provocation by the UN Secretary-General, Kofi Annan. The same challenged 50 CEOs of large financial institutions on how to integrate social, environmental, and governance factors in the capital market, in a publication of the Global Compact in partnership with the World Bank, called Who Cares Wins.

At the same time, the sustainability agendas of various non-governmental bodies (the UN Sustainable Development Goals, for example) began to bring pressure to bear on positioning from companies and governments. In this sense, each continent began to take an appropriate stance with the urgency of adapting to a new reality of future scarcity and risks that have never been faced before. The concept of sustainability in the business environment has never become more necessary. After all, *"sustainability is the search for a balance between the supply of human needs and the preservation of natural resources, without compromising future generations"* (Brundtland Report 1987).

Therefore, ESG configures a set of factors to be analyzed with a focus on the sustainability of the business, both from a socio-environmental point of view, as well as the alignment of purpose and good management of the companies. The applications are many and will be dealt with throughout this document. However, one thing is certain. After 2020 and the diverse impacts of the pandemic on society, ESG criteria became extremely relevant points of attention in the business environment.

Being in a solid company, with impeccable governance, above-average employee retention levels, employees aligned with the company's purpose, suppliers that can supply the raw material of their products for decades, and satisfied customers with the products and good company practices can configure several advantages for the same. Both qualitatively and quantitatively. The famous fortress moat, described by Warren Buffet, when describes the ability of a business to maintain its competitive advantages to protect its long-term profits and market share from competitors.

Questions for Reflection and Study

Given the ESG aspects addressed, how can this new line of thinking be used by organizations as a means of risk mitigation?

How can the social part of the ESG aspect promote value creation in the organization?

Faced with the organization's strategic issues, how does the ESG concept fit into the formation of possible competition scenarios?

Cases to be Resolved

An organization intends to develop a strategy based on the ESG premise to improve its market performance. In this sense, propose factors that can help the manager in the field of opportunity and threat, considering the Environmental, Social, and Governance perspectives.

A financial institution intends to reduce its environmental footprint to improve its ESG performance. It is worth mentioning that this type of institution provides financial resources to companies from different sectors to promote the development of new businesses. In this sense, what precautions should this financial institution take so that capital is invested in an energy company?

References

Abosede SA (2020) Impact of indebtedness on financial performance of quoted downstream oil and gas companies in Nigeria. Int J Acad Res Account Financ Manag Sci

Abbott KW, Genschel P, Snidal D, Zangl B (2015) Orchestration: global governance through intermediaries. In: Abbott KW, Genschel P, Snidal D, Zangl B (eds) International organizations as orchestrators. Cambridge University Press, Cambridge

Burt RS (1992) Structural holes: the social structure of competition. Harvard University Press, Cambridge, MA

Bauer R, Koedijk K, Otten R (2005) International evidence on ethical mutual fund performance and investment style. J Bank Finance 29:1751–1767

Bauer R, Otten R, Rad AT (2006) Ethical investing in Australia: is there a financial penalty? Pac Basin Financ J 14:33–48

Bernhagen P, Mitchell NJ (2010) The private provision of public goods: corporate commitments and the United Nations Global Compact. Int Stud Quart 54(4):1175–1187

Brundtland (1987) Report of the world commission on environment and development. https://www.are.admin.ch/are/en/home/media/publications/sustainable-development/brundtland-report.html.

Bollen NP (2007) Mutual fund attributes and investor behavior. J Financ Quant Anal 42:683–708

Blair MM, Wallman SMH (2001) Unseen wealth—report of the Brookings task force on intangibles. The Brookings Institution, Washington, DC

Bouman EA, Lindstad E, Rialland AI, Strømman AH (2017) State-of-the-art technologies, measures, and potential for reducing GHG emissions from shipping—a review. Transp Res Part D 52:408–421

Casadesus-Masanell R, Ricart JE (2011) How to design a winning business model. harvard business review. January–February 2011. Available online: https://hbr.org/2011/01/how-to-design-a-winning-business-model (Accessed on 15 November 2020)

Capelle-Blancard G, Monjon S (2012) Trends in the literature on socially responsible investment: looking for the keys under the lamppost. Bus Ethics: A Eur Rev 21:239–250

Cortez MC, Silva F, Areal N (2009) The performance of European socially responsible funds. J Bus Ethics 87:573–588

Cummings LS (2000) The financial performance of ethical investment trusts: an Australian perspective. J Bus Ethics 25:79–92

Cook ML, Burress MJ (2013) The impact of CEO tenure on cooperative governance. Manag Decis Econ 34:218–229

Clementino E, Perkins R (2021) How do companies respond to environmental, social and governance (ESG) ratings? Evidence from Italy. J Bus Ethics 171(2):379–397. Dordrecht [u.a.] Springer Science + Business Media B.V, ISSN 1573-0697, ZDB-ID 1478688-6

Clementino E, Perkins R (2021) How do companies respond to environmental, social and governance (ESG) ratings? Evidence Italy. J Bus Ethics 171:379–397. https://doi.org/10.1007/s10551-020-04441-4

Clementino E, Richard P (2021) How do companies respond to environmental. Soc Governance (ESG) ratings? Evidence from Italy. J Bus Ethics 171:379–397. https://doi.org/10.1007/s10551-020-04441-4

Damodaran A (2021) Doing good or sounding good? A skeptical look at ESG. Disponível em: http://aswathdamodaran.blogspot.com/2020/09/sounding-good-or-ng-good-skeptical.html. Acesso em: 10 Sept 2021

Daugaard D (2019) Emerging new themes in environmental, social and governance investing: a systematic literature review. Account Fine. https://doi.org/10.1111/acfi.12479

Dhaliwal DS, Li OZ, Tsang A, Yang YG (2011) Voluntary nonfinancial disclosure and the cost of equity capital: the initiation of corporate social responsibility reporting. Account Rev 86:59–100

Ding A, Daugaard D, Linnenluecke MK (2020) The future trajectory for environmental finance: planetary boundaries and environmental, social and governance analysis. Account Fin 60(1):3–14. Available at SSRN: https://ssrn.com/abstract=3556767 or https://doi.org/10.1111/acfi.12599

DEFRA (2019) United Kingdom department of environment, food & rural affairs greenhouse gas reporting—Conversion factors 2016. https://www.gov.uk/government/publications/greenhouse-gas-reporting-conversion-factors-2016 (Last date of access 29 June 2020)

Di Lullo G, Zhang H, Kumar A (2016) Evaluation of uncertainty in the well-to-tank and combustion greenhouse gas emissions of various transportation fuels. Appl Energy. https://doi.org/10.1016/j.apenergy.2016.10.027

EY (2020) How will ESG performance shape your future? Ey 2020 (July):1–19. https://assets.ey.com/content/dam/ey-sites/ey-com/en_gl/topics/assurance/assurance-pdfs/ey-global-institutional-investor-survey-2020.pdf

Eduardo Bueno M, Salmador P, Longo-Somoza M (2014) Advances in the identification and measurement of intellectual capital and future developments in the intellectual capital research agenda : experience of the intellectus model and proposal of a synthetic index. Knowledge Management Research and Practice KMRP 12(3):339–349. ISSN 1477-8238, ZDB-ID 2116287-6

Energy R (2014) Ownership strategy. Available online: https://www.or.is/en/about-or/organization-and-corporate-governance/owners-policy/ (Accessed on 2 Nov 2019)

Edwards R et al (2014) Well-to-wheels report version 4. a JEC Well-to-wheels analysis. Joint Res Center

El-Houjeiri HM, Brandt AR, Duffy JE (2013) Open-source LCA tool for estimating greenhouse gas emissions from crude oil production using field characteristics. Environ Sci Technol 47:5998–6006

EPA (2014) United states environmental protection agency. Emission factors for greenhouse gas inventories. 1–5. https://www.epa.gov/sites/default/files/2018-03/documents/emission-factors_mar_2018_0.pdf (Last date of access: 22 June 2020)

ETC Energy Transitions Commission (2018) Reaching zero carbon emissions from shipping. https://www.ieta.org/resources/COP24/Misc%20Media%20Files/Dec7/SE16%20(3).pdf (Last date of access: 22 June 2020)

Fernández-Pérez de la Lastra S, García-Carbonell N, Martín-Alcázar F, Sánchez-Gardey G (2017) Building ambidextrous organizations through intellectual capital: a proposal for a multilevel model. Intang Cap 13(3):668–693

Friede G, Busch T, Bassen A (2015a) ESG and financial performance: aggregated evidence from more than 2000 empirical studies. J Sustainable Fin Invest 5:210–233

Friede G, Busch T, Bassen A (2015b) ESG and financial performance: aggregated evidence from more than 2000 empirical studies. J Sustain Financ Invest. 5(4):210–233. https://doi.org/10.1080/20430795.2015.1118917

García-Merino JD, García-Zambrano L, Rodriguez-Castellanos A (2014) Impact of relational capital on business value. J Inf Knowl Manage

Graham ER, Thompson A (2015) Efficient orchestration? The Global Environmental Facility in the governance of climate adaptation. In: Abbott KW, Genschel P, Snidal D, Zangl B (eds) International organizations as orchestrators. CUP, Cambridge

Greene S, Lewis A (2019) Global logistics emissions council framework for logistics emissions accounting and reporting, v2.0. Smart Freight Center

Goldreyer EF, Diltz JD (1999) The performance of socially responsible mutual funds: incorporating sociopolitical information in portfolio selection. Manag Financ 25:23–36

Guerard JB (1997) Is there a cost to being socially responsible in investing? J Forecast 16:475–490

Global Reporting Initiative (GRI) (2011) G3 sustainability reporting guidelines: reporting principles and standard disclosures. Global Reporting Initiative, Amsterdam

Global Reporting Initiative (GRI) (2013) G4 sustainability reporting guidelines: reporting principles and standard disclosures. Global Reporting Initiative, Amsterdam

Greene S, Jia H (2020) Rubio-domingo, G. Well-to-tank carbon emissions from crude oil maritime transportation/2020/; Transportation Research Part D 2020, v. 88. https://doi.org/10.1016/j.trd.2020.102587

Greene S, Jia H, Rubio-Domingo G (2020) Well-to-tank carbon emissions from crude oil maritime transportation. Transp Res Part D Transp Environ 88. https://doi.org/10.1016/j.trd.2020.102587

Gu F, Lev B (2001) Intangible assets— measurement, drivers, usefulness", working paper, Boston University and New York University, New York, NY.

Ghoul SE, Guedhami O, Kwok CCY, Mishra D (2011) Does corporate social responsibility affect the cost of capital? J Bank Financ 35:2388–2406

Goss A, Roberts G (2011) The impact of corporate social responsibility on the cost of bank loans. J Bank Fin 35:1794–1810

Greene S, Jia H, Rubio-Domingo G (2020b) Well-to-tank carbon emissions from crude oil maritime transportation. Published by Elsevier Ltd., Transportation Research Part D

Hawley JP, Williams AT (2000) The rise of fiduciary capitalism: how institutional investors can make corporate America more democratic. University of Pennsylvania Press, Philadelphia, PA
Hawley J, Williams A (2007) Universal owners: challenges and opportunities. Corporate Governance: an Int Rev 15(3):415–420
Hale T, Roger C (2014) Orchestration and transnational climate governance. Rev. Int. Organiz. 9:59–82
Hong H, Kostovetsky L (2012) Red and blue investing: values and finance. J Financ Econ 103:1–19
Hoepner AGF, Majoch AAA, Zhou XY (2021) Does an asset owner's institutional setting influence its decision to sign the principles for responsible investment? J Bus Ethics 168:389–414. https://doi.org/10.1007/s10551-019-04191-y
ITF (2019) International transport forum and Organization for economic co-operation and development. ITF Transport Outlook 2019. OCEDiLibrary https://www.oecd-ilibrary.org/transport/itf-transport-outlook-2019_transp_outlook-en-2019-en
Jia H (2018) Crude oil trade and green shipping choices. Transp Res Part D Transp Environ 65:618–634
Jonsdottir GE, Sigurjonsson TO, Alavi AR, Mitchell J (2021) Applying responsible ownership to advance SDGs and the ESG framework, resulting in the issuance of green bonds. Sustainability 13:7331
Kang S-C, e Snell SA (2009) Intellectual capital architectures and ambidextrous learning: a framework for human resource management. J Manage Stud 46(1):65–92
Knizhnikov A, Shvarts E, Ametistova L, Pakhalov A, Rozhkova N, Yudaeva D (2021) Environmental transparency of Russian mining and metal companies: evidence from independent ranking system. Extr Ind Soc 8(3):100937. https://doi.org/10.1016/j.exis.2021.100937
Koh P-S, Qian C, Wang H (2013) Firm litigation risk and the insurance value of corporate social performance. Strat. Manag. J. 35:1464–1482
Lee KH, Cin BC, Lee EY (2016) Environmental responsibility and firm performance: the application of an environmental, social and governance model. Bus Strateg Environ 25(1):40–53. https://doi.org/10.1002/bse.1855
Lokuwaduge DS, Chitra S, de Silva, K (2020) Emerging corporate disclosure of environmental social and governance (ESG) risks: an australian study, Australasian accounting. Bus Fin J 14(2):35–50. https://doi.org/10.14453/aabfj.v14i2.4
Lo KY, Kwan CL (2017) The effect of environmental, social, governance and sustainability initiatives on stock value—Examining market response to initiatives Undertaken by Listed Companies. Corp Soc Responsib Environ Manag
López-Toro AA, Sánchez-Teba EM, Benítez-Márquez MD, Rodríguez-Fernández M (2021) Influence of ESGC indicators on financial performance of listed pharmaceutical companies. Int J Environ Res Public Health 18:4556. https://doi.org/10.3390/ijerph18094556
Lokuwaduge CSDS, Heenetigala K (2017) Integrating environmental, social and governance (ESG) disclosure for a sustainable development: an Australian study. Bus Strateg Environ 26(4):438–450. https://doi.org/10.1002/bse.1927
Masnadi MS, El-Houjeiri HM, Schunack D et al (2018) Well-to-refinery emissions and net-energy analysis of China's crude-oil supply. Nat Energy 3:220–226. https://doi.org/10.1038/s41560-018-0090-7
Meador JE, O'Brien DJ, Cook ML, Grothe G, Werner L, Diang'a D, Savoie RM (2016) Building sustainable smallholder cooperatives in emerging market economies: findings from a five-year project in Kenya. Sustainability (Switzerland)
Minor D, Morgan J (2011) CSR as reputation insurance: Primum non Nocere. Calif Manag Rev 53:40–59
Mishra RK, Kapil S (2016) Study on corporate governance mechanisms. Int J INDIAN Cult Bus Manag. 12(2):179–203. https://doi.org/10.1504/IJICBM.2016.074480
MERITUM (2002) Guidelines for managing and reporting on intangibles, TSER Programme, MERITUM, Tucson, AZ

Monks RA (2001) The new global investors: how shareowners can unlock sustainable prosperity worldwide. Capstone, Mankato
Nahapiet J, Ghoshal S (1998) Social capital, intellectual capital, and the organizational advantage. Acad Manage Rev 23(2):242–266
Nonaka I, Takeuchi H (1995) The knowledge-creating company. Oxford University Press, Oxford
Oprean-Stan C, Oncioiu I, Iuga IC, Stan S (2020) Impact of sustainability reporting and inadequate management of esg factors on corporate performance and sustainable growth. Sustain 12(20):1–31. https://doi.org/10.3390/su12208536
Olmer N, Comer B, Roy B, Mao X, Rutherford D (2017) Greenhouse gas emissions from global shipping, 2013–2015. https://theicct.org/publications/GHG-emissions-global-shipping-2013-2015 (Last date of access 22 June 2020)
Ozawa A, Inoue M, Kitagawa N, Muramatsu R, Anzai Y, Genchi Y, Kudoh Y (2017) Assessing uncertainties of well-to-tank greenhouse gas emissions from hydrogen supply chains. Sustainability 9:1101
Olson M (1971) The logic of collective action: public goods and the theory of groups. Harvard University Press, Cambridge, MA, USA
Pandey S (2015) Third IMO GHG study 2014. International Maritime Organization (IMO) London, UK
Loop P, DeSmith S, McNaughton B, Carey C (2020) Mind the gap: the continued divide between investors and corporates on ESG. Gov Insights Cent PwC's ESG Pulse
Petzick JA, Scherer RF, Biodzinski JD, Quinn JF, Ainina MF (1999) Global leadership skills and reputational capital: intangible resources for sustainable competitive advantage. Acad Manag Exec 13(1):58–68
Pechlaner H, Bachinger M (2014) Regional core competencies as a basis for entrepreneurship: the german hop-growing area of the Hallertau. Int J Entrepreneurship Innovat 15(1):41–50
Poulsen RT, Ponte S, Lister J (2016) Buyer-driven greening? Cargo-owners and environmental upgrading in maritime shipping. Geoforum. https://doi.org/10.1016/j.geoforum.2015.11.018
Rahman MM, Canter C, Kumar A (2015) Well-to-wheel life cycle assessment of transportation fuels derived from different North American conventional crudes. Appl Energy. https://doi.org/10.1016/j.apenergy.2015.07.004
Raza Z (2020) Effects of regulation-driven green innovations on short sea shipping's environmental and economic performance. Transp Res Part D: Transp Environ 84:102340
Mishra RK, Kapil S (2016) Study on corporate governance mechanisms. Int J Ind Cul Bus Manage 12(2)
Renneboog L, Ter Horst J, Zhang CD (2008a) Socially responsible investments: institutional aspects, performance, and investor behavior. J Bank Finance 32:1723–1742
Renneboog L, Ter Horst J, Zhang C (2008b) The price of ethics and stakeholder governance: the performance of socially responsible mutual funds. J Corp Finan 14:302–322
Roos G, Roos J (1997) Measuring your company's intellectual performance. Long Range Planning 30(3):413–426+325
Shvarts E, Pakhalov A, Knizhnikov A, Ametistova L (2018) Environmental rating of oil and gas companies in Russia: how assessment afects environmental transparency and performance. Bus Strateg Environ 27(7):1023–1038
Soderstrom N (2013) Sustainability reporting: past, present, and trends for the future. Insights Melburrne Bus Econom 13:31–37.16: 278–288. https://doi.org/10.1080/0969160X.2004.9651716
Sjöström E, Welford R (2009) Facilitators and impediments for socially responsible investment: a study of Hong Kong. Corp Soc Resp Environ Manage https://doi.org/10.1002/csr.210
Subramaniam M, Youndt MA (2005) The influence of intellectual capital on the types of innovative capabilities. Acad Manag J 48(3):450–463
Sparkes R (2002) Socially responsible investment: a global revolution. Wiley, New York
Statman M (2000) Socially responsible mutual funds (corrected). Financ Anal J 56:30–39
SBT (2018) Science-based targets inititative. Transp Sci -Based Target Setting

Schleifer P (2013) Orchestrating sustainability: the case of European Union biofuel governance. Regul Govern 7(4):533–546

Smith TWP, Jalkanen JP, Anderson BA, Corbett JJ, Faber J, Hanayama S, O'Keeffe E, Parker S, Johansson L, Aldous L, Raucci C, Traut M, Ettinger S, Nelissen D, Lee DS, Ng S, Agrawal A, Winebrake Hoen JJ, Chesworth M (2012) United nations secretary-general's high-level panel on global sustainability. Resilient People, Resilient Planet: A future worth choosing. (.pdf). New York: United Nations. p 24. Consultado em 30 de janeiro de

The Global Compact (2004) Who cares wins: connecting financial markets to a changing world. *Who Cares Wins Connecting Financial Markets to a Changing World*, 1–59. Retrieved from https://www.unepfi.org/fileadmin/events/2004/stocks/who_cares_wins_global_compact_2004.pdf

Winebrake JJ, Corbett JJ, Meyer PE (2007) Energy use and emissions from marine vessels: a total fuel life cycle approach. J Air Waste Manage Assoc 57(1):102–110

Whelan T, Atz U, Von Holt T, Clark C (2015) ESG and financial performance: uncovering the relationship by aggregating evidence from 1000 plus studies published between 2015–2020. New York University Stern, Center for sustainable Business. New York

Julio Vieira Neto has D.Sc. in Civil Engineering at Fluminense Federal University, master's in management system at Fluminense Federal University, Postgraduate with MBA in Organization and Strategy, and MBA in Business Management and Graduate in Business Administration. He is currently Associate Professor at Fluminense Federal University in the Department of Actuarial Sciences and Finance where he teaches the subjects: Demographic dynamics and quantitative methods. In the Stricto Sensu Courses, he teaches the following subjects: Life cycle analysis, integrated reporting, bibliometrics, and quali-quantitative methodology. Researcher of the Doctoral Program in Sustainable Management Systems and the Professional master's degree in management systems, both at UFF, in the areas of product life cycle management, product life cycle cost (CCVP), intangible assets and integrated reporting, topics related to organizational sustainability. He serves as Editor-in-Chief of the Brazilian Journal of Operations & Production Management (BJO&PM - ISSN (Online): 2237–8960)

Luiz Perez Zotes graduated in Civil Engineering from the Pontifical Catholic University of Rio de Janeiro (1973), with a master's degree in civil engineering from the Fluminense Federal University (1992) and a PhD in Production Engineering from the Federal University of Rio de Janeiro (1994). He is currently Full Professor at the Fluminense Federal University. He has experience in the areas of general administration, with an emphasis on financial administration, strategic management and project management, as well as in areas of civil engineering and security.

Flávio Ezequiel De Andrade Ph.D. student in Production Engineering (UFF), master's in management systems (UFF), MBA in Business Management (Ibmec-RJ), Postgraduate in Analysis, Design, and Systems Management (PUC-RJ), and Graduate in Statistics (ENCE). I have been working at Globo since 2014, currently working as Agile Operations Manager, works as Researcher (Research Groups CNPq–UFF) and Professor of Project Management in the MBA in Strategic Business Management (UFF). I built my career working in organizations of great relevance in the national/ international scenario in the segments of media, mining, telecommunications, and call center, exercising functions in areas of management, quality, finance, processes, statistics, projects, and market research. Currently, I'm interested in studying ESG in its social pillar, specifically racial diversity.

Adriane Domingues Quelhas graduated in Architecture and Urbanism at Santa Úrsula University (1996), Specialization in Occupational Safety Engineering at Fluminense Federal University (1997), master's in management system, and D. Sc in Civil Engineering at Fluminense Federal University (2017) and Post-Doctorate in Sustainable Management System at Fluminense Federal

University (2019). His line of research is in the area of organizational behavior with emphasis on Organizational Culture and Climate (cognitive and emotional aspects) and the impacts on the work environment and organizational performance. She taught classes as a contract professor at the State University of Rio de Janeiro (UERJ) and as a guest professor at the Getúlio Vargas Foundation (FGV). She currently teaches classes as a guest professor at the Fluminense Federal University in Stricto Senso and Lato Sensu courses in the areas of culture and organizational climate, strategic people management, workplace safety, and scientific methodology.

Nathália De Almeida Marques has Master's in business management (UFF), Post-Graduation in Safety Engineering (UFRJ) and Graduation in Chemical Engineering (IME), Extension courses in Renewable Energies (PUC-SP) and Strategic Carbon Management (FGV-RJ). PMP (PMI) and ITIL 4 Foundation (Axelos) certifications. Currently Nathália De Almeida Marques works as Project Coordinator (Wood) and Researcher in the area of Social Capital (Research Groups CNPq–UFF), been working for years in multinationals on oil and gas sector, and also having experiences in the pharmaceutical, technological (fintech) and media segments, performing functions on project management area (predictive and agile) of the most diverse disciplines and magnitudes.

Bernardo Shogo Okada Ahmed Federal Fluminense University's Undergraduate Student, Minor in Entrepreneurship and Innovation, with Corporate Finance underground by COPPEAD UFRJ. Bernardo has passed by companies like J. Safra Bank and Turim Multi Family Office, where it held the responsibility of integrating the Sustainable Investing Strategies within the Investments Team.

Chapter 3
Regional Innovation Factors and the Future of Work: The Contribution of ESG Concepts

Michelle Do Carmo Vieira and Marcio Silva Borges

Abstract Current transformations highlight the role of innovation in the performance of organizations and in the future of work. This demonstrates the importance of companies making innovation compatible with better environmental, social, and governance practices, known as ESG, since this relationship influences stakeholders and the territory. This chapter seeks to present the regional factors of innovation and some perspectives on the future of work, emphasizing the importance of ESG dimensions for understanding the role that organizations can play in social development and the territories in which they operate. Using case studies with ESG experiences, reflections, and cases to be solved, the aim is to demonstrate that the efficient management of resources, innovation and investment in human capital are key to boosting the competitiveness and success of organizations.

Keywords Innovation · ESG · Future of work · Competitiveness · Stakeholders

3.1 Introduction

The current moment is marked by the speed with which transformations are taking place in almost all sectors of society. At the center of these changes, innovation appears as a fundamental element for the survival and success of organizations, providing highly competitive circumstances, leading them to seek new management models and outline competitiveness strategies. The successful implementation of ESG Environmental, Social and Governance management requires a holistic

M. Do Carmo Vieira (✉)
Territorial Development and Public Policy, Program for Territorial Development and Public Policy (PPGDT), Federal Rural University of Rio de Janeiro (UFRRJ), Seropédica, Brazil
e-mail: michellemcv@ufrrj.br

M. S. Borges
Technology and Innovation Sciences in Agribusiness, Program for Territorial Development and Public Policy (PPGDT), Federal Rural University of Rio de Janeiro (UFRRJ), Seropédica, Brazil
e-mail: marcioborges@ufrrj.br

M. J. Meiriño and O. L. G. Quelhas (eds.), *Critical Factors Associated with Environmental, Social and Governance Management in Organizations*, Accounting, Finance, Sustainability, Governance & Fraud: Theory and Application, https://doi.org/10.1007/978-981-96-1725-8_3

approach, the identification of critical factors and commitment at all organizational levels, benefiting from competitive advantages and positive recognition from stakeholders.

As the capacity for innovation boosts the competitiveness of geographic areas seeking to solidify their relevance on the global stage, it is also reshaping the future of work. What emerges from this is a transition in labor relations, motivated by the evolution of production processes, in which intellectual capital acquires an increasingly greater role. In the face of technological advances and new management models, the capacity for continuous learning, creativity, flexibility, and adaptability become essential skills for workers, as organizations seek innovation and solutions to complex problems.

The future of work in the context of ESG factors implies a shift towards more ethical, sustainable, and socially responsible work practices. Organizations that adopt these principles are better positioned to attract and retain talent, earn the trust of stakeholders, and become increasingly valued. In this chapter, we will look at the complex dynamics that connect these three essential factors, examining how innovation can contribute to regional development and shape the future of work, while also considering the implications for society.

3.2 Role of Chapter in the Structure of the Book

In a world where innovation, good governance practices, and social and environmental responsibility have become imperative for the success and sustainability of organizations, these can no longer disregard the risks and potential impact of their operations over territories. To insert the social and territorial perspective in the development strategy and business is a pathway for transforming investments into conditions to improve the quality of life of society and the organization of work, reinforcing values and maximizing opportunities and results.

In a context of regional economic development and growth, innovation is a decisive factor for strengthening the competitiveness of local producers, managers, and entrepreneurs, resulting in positive effects on economies and regional spaces. The promoting agent of these transformations is no longer restricted to capital, as it now involves knowledge, new ideas, skills, creativity and human capital, shedding light on the importance of innovation as a potential means of generating greater returns and value distribution in territories.

It is due to this dynamic scenario of transformations that this chapter intends to present the regional factors of innovation and some perspectives on the future of work, emphasizing the importance of the ESG dimensions addressed in this book, for the understanding of the social role of the organizations where these may play a part.

3.3 Theme Development History

Since the 1990s, competitiveness criteria have undergone new changes to the detriment of global dynamics. If the focus had previously been on production, in which factors such as quality control and low cost were preponderant for obtaining competitive advantages, nowadays these requirements have become the bare minimum for access to the most diverse markets. Thus, it can be said that the main challenge for companies to remain in leadership is centered on the ability to develop new technologies, access new markets and management methods, as well as the integration of value chains that include social responsibility and sustainability. Therefore, flexibility, speed, innovation, and integration are essential elements for competitiveness. (Hitt et al. 2017; Kotler and Keller 2018).

The structuring of the business strategy should be thought of in terms of how to achieve and sustain competitive advantage and how to effectively carry out organizational changes. Companies are only able to overcome its rivals if they are able to set themselves apart and sustain that position by providing greater value to customers or creating comparable value at lower costs, or both (Porter 1996, p. 62).

Alves and Costa (2020, p. 954) suggest that to create a competitive advantage, companies must find a way to differentiate themselves from their competitors. They show that this can be achieved through marked placement strategies, by supplying unique products and services or creating dynamic resources and unique and complementary opportunities to respond to rapidly changing market conditions. To gain a competitive advantage, companies must primarily be an aid to customers in their journey of discovering what it is that they want (Kotler and Keller 2018, p. 9).

Factors such as the proximity of raw material sources and cheap labor no longer serve as competitive differentiating factors in open markets and the relentless pursuit of quality and flexibility by companies. The advances provided by applied technology and industry have gradually reduced the competitive advantages provided by natural resources. On the other hand, rising competition has led to the diversification of products in the face of an increasingly sophisticated and eager-for-quality market. In this regard, the low cost of labor cannot be pointed out as a competitive advantage for companies, since in most markets' competition has been focused on other issues not directly linked to the price of the product, but rather its quality, technology and cultural appeal (Borges 2014).

Innovation has been a constant in human development. From the discovery and mastery of fire to the creation of electricity, a series of transformations have significantly changed how we work in terms of behavior and territorial intervention, conditioning new modes of production and consumption in modern society.

The ability to develop new products and services, improve their processes, establish new markets, provide rapid responses to customers and master new technologies, all characterized organizations that understand the relevance of innovation and knowledge to drive their results and sustain their long-term performance (Hall and Sena 2017; Li et al. 2017).

One of the main theorists to recognize the influence of innovation as a driver of development in the capitalist economy was the Austrian economist Joseph Alois Schumpeter (1883–1950). His conceptions highlight the role of companies and entrepreneurial individuals as fundamental elements of the innovation process. For Schumpeter (1982), given the current dynamism present in society, the capitalist economy moves through cycles of peak and crisis, producing regular movements through a structured pattern of expansion and retraction. The development process takes place through endogenous transformations to the economic system, motivated by innovation. This process then initiates a new economic cycle that causes changes in industrial organization with the introduction of new methods, technologies, processes, opening of new markets, among other aspects.

Considering this dynamic characteristic of the capitalist system, there can only be innovation if there is creative destruction, a fundamental problem of capitalism, which moves forward by destroying old elements and incessantly creating new ones. This process of creative destruction is basic to understanding capitalism. It is from this that capitalism is constituted and the whole capitalist enterprise must adapt to it in order to survive (Schumpeter 1984, p. 111).

The speed of the creation of economic goods and their diffusion is what differentiates the innovation process that we are experiencing today from those that took place throughout past historical processes of technical evolution (Tunis 2020, p. 59). The positive potentials now attributed to the new cycle of innovation evoke and expand those attributed to the previous waves of innovation linked to ICT technologies, and even to the transition from Fordism to post-Fordism (Caruso 2018).

For Barbosa et al. (2020, p. 12), it was during the first Industrial Revolution that science began to be assimilated as a key to solving problems in production and product development. Thus, to remain profitable and competitive, since the late nineteenth century companies have implemented research and development (R&D) departments. However, companies' access to new technologies is often limited by patents, which unbalances the significant economic advantages of innovation and makes competition with other companies unequal.

The history of the world economy since the Industrial Revolution has been marked by intense technical progress, continuous but irregular economic growth, increasing globalization, and an increasingly complex world division of labor (Hobsbawm 1995). The study of industrial progress and its effects is hardly young, as well as its attempt to shape the worker "of the future" and the cities they occupy. In this context, many of the plans recognize that the true disruptive effect of new technologies on the future of work is not yet known, especially in terms of jobs that will be extinguished, created or displaced (Barbosa et al. 2020).

The concern with the effects of technology on conditions and jobs has accompanied society over time, giving rise to labor movements, such as Luddism, which took place in England during the nineteenth century (Piva and Vivarelli, 2018; Gera and Singh 2019). Thus, it was not technical progress that workers riled against, but any threat to changes in social relations of production (Frey and Osborne 2017). Since such relations became guided by market logic and productivity, the proletarianization and precariousness of labor relations has become more evident.

The accelerated pace of production and the renewal of knowledge subject them to rapid obsolescence, reflecting the nature of the worker in the current society, which consists in learning, transmitting knowledge and producing knowledge (Lévy 1993). Further studies have also pointed out that, for innovation to be a continuous process, managers need to focus not only on products, technology and processes, but also on the organizational culture, norms and values that govern the organization (Matinaro and Liu 2017; Axe and Greve 2017).

Innovation has had a significant impact on the labour market. The restructuring of organizations, the emergence of new roles, the demand for new skills and the adoption of new behaviors, for example, brought about many challenges for professionals and companies in the twenty-first century. The impact of these innovations is not only restricted to manual and routine tasks, but also to cognitive and unusual ones. This puts workers at risk due to the automation and digitization processes of a large part of organizational operations (Schwab 2018; Novakova 2020).

The future of the labor market was also analyzed by Hawksworth et al. (2018). The authors found that in the long-term perspective, less educated workers are particularly exposed to the risks of automation, in a future that prioritizes increasing investment in lifelong learning and refreshing. This only stresses the need to invest in qualifications for the development of new skills.

Given this scenario, sustainability and social issues have been gaining more and more space due to the new dynamics produced, thereby constituting a growing demand in the labor market. A significant aspect of organizational development and maximizing its results stems from its purpose and its ability to promote stakeholder engagement. This will reflect the way the organization relates to society, which can provoke positive responses and lead to the obtaining of heterogeneous resources in a way that positively enhances their innovative capabilities (Taliente et al. 2019).

3.4 Conceptual Framework—Innovation, Competitiveness, the Future of Work, and ESG

Knowledge is the basis of all innovation and both are indispensable factors for the development of science, technology, models and processes, all in constant transformation. Innovation is directly associated with the creation of something new, which van be goods produced by the industry or a service provided to the market in an original manner (Tunes 2020, p. 17).

Innovation is more than a new idea or an invention. An innovation requires implementation, whether it is put into active use or made available for use by other parties, companies, individuals or organizations (OECD 2018, p. 44). It is part of a dynamic process that contributes to organizations developing new skills, raising quality and enhancing results. The application of knowledge in the organization has transformed it into an asset necessary for innovation, differentiating this economic resource from other factors of production.

In the labor market, there has been an increasing polarization of opportunities between high and low skill paying jobs, as well as unemployment and underemployment, income stagnation and income inequality (Antunes 2006). On the other hand, there is an expansion in the development of information and communication technologies, artificial intelligence, microelectronics, robotics, nanotechnology, among other cutting-edge sectors, evidencing a wide possibility of scenarios that have been subjected to automation. The new jobs emerging in this universe will require specific technical and complex problem-solving skills and are therefore not accessible to most workers.

Data obtained from the Future Jobs Report presented at the World Economic Forum 2020 indicated that in the last two years there has been an increase in the use of technologies by the companies studied. The theme of the forum focused on the current situation and the future conditions of work in relation to the social-economic scenario and the impacts of digital technologies that suppress jobs. According to the research, the transformations related to the future of work will be guided by a set of new technologies, sectors, markets, and economic systems (Fig. 3.1). An imminent possibility of displacement of the workforce was observed, given the scarcity of human skills, motivated by the increase in automation. According to the probability of companies adopting technologies by 2025, such cloud computing, big-data and e-commerce, these continue to be a high priority, as pointed out in previous years. However, there has been increased interest in encryption, which, according to the analysis, reflects new vulnerabilities in our digital age. The significant increase in the number of companies hoping to adopt non-human robots and artificial intelligence was also observed.

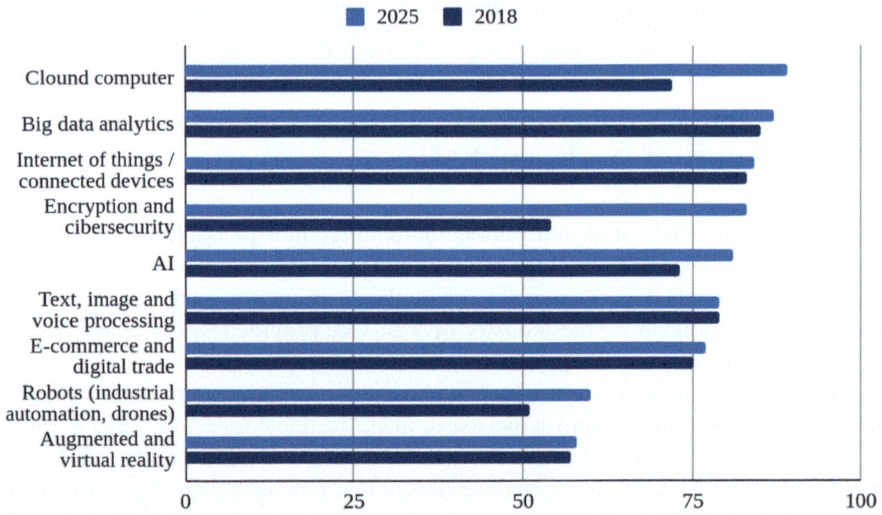

Fig. 3.1 Technologies that will probably be adopted by 2025, by the companies surveyed. *Source*: Adapted from The Future of Jobs Survey (World Economic Forum, 2020)

The report also revealed the effects produced by a new global recession in 2020 caused by the COVID-19 Pandemic, which caused changes in economies and the labor market, affecting social well-being and productivity. This reverberated in the immediate increase in unemployment, causing the displacement of workers from their positions (World Economic Forum 2020, p. 10).

The mapping of future job opportunities concluded that there is a significant likelihood of increased demand in the future for both "digital" and "human" professions. According to Kon (2020), analysts indicate that the most recent Industrial Revolution, associated with changes both demographic and in consumer needs, is creating the demand for a considerable rise in new job opportunities, in companies and beyond them, although at the expense of unequal opportunities and income. Therefore, the new professions should reflect the importance of human interaction in the new economy, especially in certain areas that require direct contact with the population.[1]

According to Harari (2018), the emergence of new forms of work and the training of individuals will be a recurring process, as will the change of employment and work, replacing the past trend of permanent and continuous employment, positions, or professions. The qualities to be expected show the search for constant adaptation skills, given the speed of transformations.

The challenges posed by the process of globalization and structural changes in the world economy also produced effects that demand analysis on the insertion of territorial dynamics as a catalyst for investments. The perception that regional factors can stimulate innovation has led to a growing interest in the analysis of innovation at the regional level (OECD 2005, p. 48).

The role of innovation as an advantage for attracting capital can boost territorial development, which can transform certain spaces and provide improvements in quality of life. Sposito (2019, p. 33), however, stresses that there are territories with a higher density of technological activities and others with a lower technological density, which shows the regionalization aspect of technological capacity. The author argues that densities, rhythms and abilities are unevenly distributed, from the raw materials and potentialities given by natural factors to the territorial division of labor. "This causes different regional potentialities that are not restricted to infrastructure, but mainly to elements of the superstructure that accelerate or slow down uneven or integrated rhythms" (Sposito 2019, p. 33).

Tunes (2020) adds that the selectivity of knowledge that matters to capital also influences and guides investments in innovation, denoting specific forms of spatial arrangements in different regions. Thus, knowledge, research in development and technology, connectivity and information networks, productive agglomerations,

[1] Transitions to new professions will be based on the skills of current workers and their potential for future roles, but also on the requalification of the workforce to incorporate new roles, which requires government and private investment. The adaptation of physical and regulatory infrastructure, as well as the requalification of workers for the new paradigm, requires not only the perception of additional investment needs from governments and companies, but mainly partnership-based actions to drive change in legislation and business practices (Kon, 2020, p. 202-203).

intellectual capital and environments themselves, among other elements, can condition innovation at the regional level, without necessarily being included in the same type of system.

The ability to innovate is a key element for regional competitiveness. However, in order for development to materialize in these spaces, policies need to be in place, which requires the elaboration of strategies and synergies between different public and private actors in society, mainly involving human resources. These aspects have revealed new behaviors that reflect demands for better management practices, social and environmental responsibility.

The demand for sustainability has become essential in contemporary society, which has led companies to have to rethink their roles as agents of social transformation and to seek to understand how their decisions impact their surroundings. The definition of environmental, social and governance criteria, called ESG, has assumed a prominent position in organizational strategies, being increasingly perceived as a crucial factor in the conduct of companies' investment decisions through a path of sensitivity towards environmental and social issues (Brogi and Lagasio 2018).

The term Environmental, Social and Governance—ESG was first used in 2004 in a report entitled "Who cares wins" published by the World Bank in partnership with the United Nations (UN) Global Compact and has been adopted on a larger scale since 2005 (Redecker and Trindade 2021). The initiative established criteria and recommendations for the incorporation of environmental, social and governance issues in the financial and capital markets, as such practices make organizations contribute more to society and act sustainably. In general, the literature defines the ESG as an index that evaluates organizational practices for investment analysis based on the pillars of environmental, social and governance sustainability (The Global Compact 2004; Sion 2021; Guerra Filho Turqueti and Lima 2021).

A good ESG score reveals a company's commitment to their social and environmental responsibilities, which decreases asymmetric information and stock price volatility in the market (Lueg et al. 2019; Shakil 2020).

The question that arises then is, to what extent social and environmental issues affect the competitiveness of organizations, and whether, from the point of view of governance, companies are actually concerned with the issues or are only using them as marketing tools for improving their image to leverage their business.

Much of the problem lies in the companies themselves, which remain attached to an outdated value creation approach that has emerged in recent decades (Porter and Kramer 2011). These companies, the authors claim, continue to have a short-sighted view of value creation, optimizing short-term financial performance while ignoring the most important customer needs and major factors that determine their long-term success.

What is expected from the adoption of ESG practices, as a principle for driving investment, is that they promote improvement in the environmental, social, and economic performance of organizations and that the movement towards this principle is considered an opportunity, rather than an obligation to be met. Porter, Serafeim and Kramer (2019) suggest that there is a new stage of understanding in the link between investment performance and social impact. They point out that previous approaches, such as socially responsible investment and environmental, social and governance assessment, have obscured the opportunities for greater growth, profitability, and competitive advantage that come from tackling social and environmental issues as an integral part of a company's central strategic placement.

Porter, Serafeim and Kramer (2019) also show that many investors select potential investments through a purely financial analysis that ignores social issues, and then use the company's overall ESG performance as a final mechanism to reduce risk. Thus, such investors assume that better scores on all ESG indicators indicate more prudent and insightful management. On the other hand, they point out that other investors completely deviate from the fundamental analysis of the company in favor of passive investment, less fundamental research, broad diversification, focus on liquidity, and trading in short-term catalysts.

Porter and Kramer (2011) demonstrate that the purpose of the company must gear towards generating shared value, not just of profit for the sake of profit, which could then fuel the next wave of innovation and productivity growth in the global economy. Shared value involves the creation of economic value to also generate value for society by tackling its needs and challenges. In that sense, it is necessary to reconnect the success of companies to social progress. It has to do with sharing the value already generated by the company (Porter and Kramer 2011).

ESG actions in this context imply in the company becoming a transformative agent in its segment and in society. An approach in which the analysis of social impact is dissociated from the analysis of competitive strategy and growth misses out on important sources of competitive advantage that come from some, albeit not all, improvements in corporate social performance (Porter et al. 2019). This requires governance aimed at developing strategies based on principles, values, ethics, transparency and responsibility and establishing integrated goals, because isolated action does create sustainability now engagement with the ESG principles in companies.

Porter and Kramer (2011) point out that "the perception that companies thrive at the expense of the surrounding community is widespread," especially due to the assumption that most companies remain entrenched in a social responsibility mentality that has social issues at the periphery of their activities, rather than their center. It is not a matter of philanthropic practice, but of recognizing and thinking about new forms of economic progress through social well-being and environmental conservation.

The social impact of companies is commonly associated with the quality of the products and/or services marketed by them, but it is much more than that. The exercise of actions with greater engagement and social responsibility enhances the value of intangible assets and contributes to the economic development of territories through a qualitative dimension that includes the reduction of inequalities and the improvement of living conditions.

It is not only aimed at economic growth in the territories where the company or organization operates, but also at the sustainability of its actions, which demonstrates a long-term organizational vision capable of conveying values and trust to its stakeholders and the market. Although environmental and social responsibility issues involve discussions that are not recent, the ESG criteria stand out by defining conditions and standards of conduct that can qualify investment opportunities with sustainability.

Regarding the themes presented so far in this chapter, it should be noted that the social dimension of ESG covers the social risks and the impact of the company's relationship with its stakeholders, that is, the effects that materialize after the delivery of the result and that can transform social and territorial settings. It is necessary to reflect on what companies are doing or can do to promote social inclusion, respect for human and workers' rights, diversity, equity, and citizenship, among other aspects that can add value to communities and promote better living conditions, in addition to maximizing results. Companies that take into account the interests and needs of society in carrying out their economic activity, and whose management seeks to drive improvements for the society, unequivocally respect and carry out their social function (Redecker and Trindade 2021).

Organizations can actively contribute to improving their surrounding territories. In the same way that one can obtain benefits from endogenous resources, social actors and existing infrastructures, companies can contribute to social-spatial development. They can favor the generation of knowledge, train the local workforce, and thus access more qualified human resources, improve and strengthen the productive structure through technological development. These spaces can be consolidated as innovative means to enhance development and add social, economic, and territorial value.

3.5 ESG Experiences in Brazil and in the World

Box 3.1 Natura and Co (Natura 2020)

The largest Brazilian multinational in the cosmetics sector, constituted by Avon, The Body Shop and Aesop, it stands out for implementing a comprehensive innovation process that supports its management culture and business model. The group linked managers' performance indicators and bonus ratings to environmental and social goals, as well as their financial results, thus reinforcing ESG initiatives.

Since 2012, Instituto Natura, in partnership with CEDAC—Educational Community, has been developing the Trilhas project, a social technology to assist the work of elementary school teachers. The project has since been integrated into policy by the Ministry of Education, and today serves approximately 3 million students in the public school system across 72.000 institutions.

Data obtained from the Natura GRI 2020 Annual Report show that Natura & Co obtained second place in the Refinitiv Diversity and Inclusion (D&I) Index – Top 100, by Refinitiv, a Reuters news agency and Blackstone fund company. In 2019, Natura ranked second place in the index. Natura is one of the 25 best companies in the world to work for, according to the World's Best Workplaces 2020 ranking. It reached first place in the sixth edition of the Valor Inovação Brasil 2020 Award, granted by the newspaper Valor Econômico.

As for social progress, in order to generate innovation and new businesses in the Amazon, maintain the forest, and improve the quality of life of traditional peoples, in December 2020, the Forest Knots alliance was held between Natura and the NGO Conexsus—Sustainable Connections, with the objective of strengthening the connections between entrepreneurs, community associations, financial institutions, companies, universities, governments, and international organizations. Some of the goals of the alliance include fostering local businesses, with emphasis on those led by women and young people; solving impasses in production chains; and increasing the revenue of community organizations and their participation in public calls.

Striving for more diversity and inclusion, 2020 ended with 51% of women in leadership positions at the three highest levels of the organization in Natura & Co Latin America. Regarding Natura, the percentage was the same, which demonstrates efforts to achieve a more balanced gender representation in the company. In relation to the goal of professionals with disabilities (PwD), the index was 7.3%, slightly below the stipulated 8%. From 2014 to 2020, this indicator increased in 46% and, in the last year, the variation was of only 0.1 percentage point when compared to 2019, a factor surely influenced by the effects of the pandemic. The group offers training journeys and workshops on self-knowledge, emotional and relational intelligence and professional protagonism.

In October 2020, Natura Saúde was launched in Brazil, a platform that expands health benefits that had already been available to consultants and leaders for a few years. The platform has three partners who offer consultations (in person or virtually) and tests with prices up to 35% cheaper than those found in the market.

As a result of the effects of the Covid-19 pandemic, the company developed an income transfer mechanism for consultants in extreme social vulnerability, through which resources were allocated according to food security criteria, level of dependence on the income generated by the consulting activity, Covid-19 symptoms and belonging to risk groups. The emergency fund benefited 3.757 Brazilian consultants and 5.260 across throughout Hispanic America. BRL 655.200 were allocated to the fund, to provide food stamps, telemedicine care and, in some cases, an income supplement that ranged from BRL 80 to BRL 220. In all, the support provided to consultants totaled about R$1 million, including actions to combat domestic violence and hygiene products.

Box 3.2 ESG Strategies for Industries (FIRJAN 2021, p. 12)

The Federation of Industries of the State of Rio de Janeiro (FIRJAN) consists of five institutions that support and encourage industries, generate ideas, jobs and develop solutions: FIRJAN, FIRJAN Senai, FIRJAN SESI, FIRJAN CIRJ, and FIRJAN IEL. Considering that industry development is fundamental for the economy and social well-being, FIRJAN operates in key areas such as competitiveness, education, and quality of life for workers and their families. They contribute to stimulate technological innovation, support initiatives that enable the full development of the state of Rio and of society as a whole.

The growing relevance of environmental, social and governance management strategies points to a necessary transformation agenda. What are the items that will increase the value of companies and ensure their reputation in the long term? Identifying these points and finding metrics to evaluate them is a strategic issue relevant to all. For this reason, FIRJAN developed research that sought to understand the current state of these factors and identify the most adopted ESG criteria in large companies, as well as those most required by these companies in their supply chains, which are usually integrated by smaller businesses. The results show that:

- Companies know and perceive positive value in the adoption of ESG criteria, so there is a tendency for the theme to become more consolidated in corporate management with time.
- Significant actions are already being developed in the context of ESG, with emphasis on multinational companies.
- ESG criteria are explored both for monitoring the company's own performance and in the management of its suppliers.
- The most reported motivation behind ESG implementation is the company's image and reputation management.
- From a supplier management perspective, the primary motivation for requiring ESG criteria is risk minimization.
- Most companies operate under the three ESG axes, but the management of the environmental axis criterion prevails.
- The governance criterion is the most relevant one when hiring suppliers.

There are still challenges for the consolidation of ESG strategies and for the understanding of the required criteria. Given this context, FIRJAN compiled a set of recommended practices and essential actions that can guide the management in each of the ESG criteria pointed out as most relevant. In addition, each criterion was associated with corresponding tools and guidelines that can be used as technical references to support the structuring of ESG strategies for industries.

Box 3.3 Investment Market (Neto 2021)

In 2019, 181 CEOs of mega corporations signed a "Statement of Purpose" document, committing to the adoption of new ESG practices attuned to sustainable development and greater corporate social responsibility. In the wake of "*green bonds*," we've seen the rise of social bonds as well in the financial market, focusing on sustainable social impact, which can be connected to the Sustainable Development Goals (SDGs) defined by the United Nations in 2015 and which would guide the 2030 Agenda for sustainable development. According to the IDB, the theme-based securities market, created just over ten years ago, has grown in line with the demands of investors, who place increasing importance on "risk management and governance" around the "social and environmental impact" generated by businesses.

As early as late 2017, French President Emmanuel Macron promoted the *One Planet* meeting in response to Donald Trump's threats to leave the Paris Agreement. At that time, eight central banks created the Network for the "Greening of the Financial Sector." "This network today contains 69 central banks and a list of 13 observers, including the IMF, World Bank and the Inter-American Bank—IBRD, as well as several regional banks" (according to Professor José Eli da Veiga, FEA/USP).

The theme-based debt securities market (or friendly bonds)—the so-called green, social or sustainable bonds, aimed especially at combating practices that cause global warming and environmental degradation—totaled US$328 billion in 2019 (R$1.8 trillion) in emissions worldwide last year, up 57% when compared to 2018 (data compiled in early June last year by the Inter-American Development Bank (IDB), based on information from the Bloomberg agency in July 2020).

In Brazil, such investments began to emerge quite timidly. According to the Brazilian Association of Financial and Capital Market Entities–Anbima), currently there is about BRL 700 million invested in ESG stock funds, which is equivalent to only 0.13% of the total investments in equity funds in Brazil. Despite this, some environmental impact investment funds have shown significant growth over the past few years. Just as an example: MOV Investimentos is a manager of social and environmental impact funds whose volume of investments had grown about 6.7 times since its creation (in 2012) on the 1st quarter of 2019. In the same vein, Vox Capital, also an impact investment fund, grew about 12 times in the same period.

Another important indicator to consider in this new context is the Corporate Sustainability Index (ISE) created in 2005 by the São Paulo Stock Exchange, as an equivalent to the Dow Jones Sustainability Index. Since its creation until June 2011, the rate of appreciation of this financial index has evolved in almost the same manner as the traditional iBovespa. However, from July 2011 to June 2019, while iBovespa appreciated in 225%, ISE appreciated in 240%, which only reinforces the perception that sustainable investments are here to stay.

Finally, it is worth noting that, despite the existence of some indices that refer to isolated aspects – such as the Corporate Governance Index (IGCT) and the Efficient Carbon Index (ICO2), in addition to the stock exchange indices—there is still no entity that is officially responsible for the ESG seal or certificate. It is, therefore, a market still to be developed by institutions and entrepreneurs.

3.6 Indication of How the Conceptual Framework Will Be Incorporated into the ESG Development, Implementation and Monitoring Process

Guiding companies towards sustainability requires important changes in the development of the social, economic and environmental pillars. It is necessary to define new visions of what social equality, environmental justice and business ethics mean. This requires a better understanding of the financial and physical forms of capital but, above all, of human, social and natural capital (Elkington 2020).

The demand for more conscious organizations implies reflecting on the impact of their operations on the environment and the society in which they operate. Meeting this demand has led companies to innovate while considering sustainability as an integral part of their business strategies. In this context, sustainable innovation can be supported by actions such as: valuing the role of governance in line with well-defined objectives and strategies that consider different scenarios; focusing on customer needs and value delivery; incorporating sustainability action into the main operational processes of the organization; employee engagement through incentives and training to encourage an innovative culture, among other measures (Porter et al. 2019).

The integration of ESG principles into companies' conduct enables improvement in their non-financial indicators, such as reputation, satisfaction, risk reduction, and delivery of value to their stakeholders. For this, it is important that first of all they are incorporated as opportunities in the promotion of social, environmental and economic well-being. Therefore, the purpose must be greater than obtaining higher financial returns, philanthropic actions, or obligations to be fulfilled; it is necessary to integrate economic development with social development. This requires the action of economic actors and policies that can identify opportunities to fulfill the needs society beyond just basic ones and recognizes the impact of social shortcomings throughout the value chain.

Henisz et al. (2019) demonstrated reasons why leading action based on the ESG principles contributes to increasing the value of the company. ESG compliance has a positive effect on revenue expansion and access to new markets. Tracking ESG performance reduces costs by decreasing energy and water consumption. By achieving greater strategic freedom, the ESG increases the chances of obtaining grants and other government support. A strong ESG proposal increases employee motivation and response to increased social credibility. In this way, ESG increases return on investment by improving capital allocation in the long run.

Creating social impact through an innovative and profitable business model reshapes the nature of competition and makes social impact a part of capitalism itself (Porter et al. 2019). A company needs a thriving community not only to generate demand for its products, but also to supply essential public assets and a favorable environment (Porter and Kramer 2011). On the other hand, the authors emphasized that a community needs prosperous companies to create jobs and income generation opportunities for its citizens. This interdependent relationship means that public policies that undermine the productivity and competitiveness of companies are self-destructive, especially in a globalized economy, in which facilities and jobs can easily just move elsewhere (Porter and Kramer 2011).

The applicability of ESG factors implies including organizational objectives that consider the implementation of actions aimed at the legitimacy of the company before society, which requires changes in business decisions are traditionally made, beginning with listening and interacting with stakeholders. And this degree of legitimacy influences the market value of the company, making it so that economic and social gains are now more interdependent than ever (Mazon and Issa 2022, p. 41).

The social perspective highlights issues related to social responsibility, diversity and inclusion, ethics, health and safety in the workplace, as well as compliance with labor standards and legislation. Several social aspects are considered for the implementation of ESG practices, such as improvements in work environment conditions, solutions for risk reduction, salary justice policy, professional training programs and appreciation of ethnic, cultural, social, and gender diversity, among others, both in the corporate environment and in communities (Engelmann and Nascimento 2021, p. 128).

Faced with these dynamic transformations, companies and workers need to prepare for new trends and challenges. According to the Future of Jobs report prepared from The World Economic Forum 2020, the competencies and skills of the workforce with the greatest potential relevance for innovation and monitoring the pace of change are: resilience; tolerance to stress and flexibility; emotional intelligence; problem solving; creativity; persuasion and negotiation; analytical thinking and innovation; leadership and cognitive flexibility.

Half (2021) points out that the trends for the coming years are not limited only to new jobs, as workers' profiles and forms of working are also changing. The author argues that the productive home office or hybrid model is already a reality for many companies, enabling more flexible environments and schedules. In addition, it also brings the ability of self-management, the increasing use of technology and the search for new skills to the forefront.

The social dimension, however, has more room for progress both in planning and in execution, especially towards generating greater positive impact, clarity in the practices to be adopted and on how they are assessed. In this context, recruitment systems should keep track of the changes in the focus on long-term employment contracts with previously defined technical skills requirements, as they will likely need to hire workers with multidisciplinary training and skills to dynamically assume new roles that are still to be created. At the same time, such professionals will be required for specific, national or globalized projects and the workplace should not necessarily be focused on the contracting company (Kon 2020, p. 203). In many cases, as pointed out in the World Economic Forum of 2020, it will be less expensive to encourage the work of the contractor to be developed in co-working spaces, or at home, through networking, flexible schedules, and autonomy.

To face this environment of uncertainty and unpredictability, education and individual knowledge training should be based not only on information, but also on creativity, flexibility and willingness to respond to new challenges, seeking to stimulate the creative potential and skills that machines cannot replace (Kon 2020).

The complexity of the issue poses many challenges for organizations and workers, which requires the development of flexible strategies and processes to adapt to these new scenarios. For certain sectors, the COVID-19 pandemic accelerated transformations that were already underway, intensified the use of technologies, drove innovation, as well as the need for qualified professionals.

These aspects demonstrate the importance of companies and organizations formulating strategies that involve investments in labor qualification, which can result in productive gains and improvements for society and their territories. It is clear how much the rapprochement and partnership between universities and companies in learning processes that lead to innovation has been present in many cases[2] (Tunes 2020, p. 378).

The search for the development of territorial competencies for the promotion of regional development requires the integration of different actors, such as companies, society and public authorities, for the elaboration of planning strategies. This requires the appreciation of territorial resources, stimulating the production of knowledge, the qualification of human resources, technological development, the generation of jobs, and other social, economic, and environmental potential.

3.7 Summary of Main Ideas and Concepts

Efficient resource management, territorial potential, investment in human capital, and the development of public policies that promote improvements in innovation processes are essential to stimulate competitiveness and development.

Competitive advantage is the attribute that places organizations ahead of their competitors in the market, making it a strong differentiating factor in achieving goals and results. It arises from the ability to deliver a higher value than others (Porter 1996; Kotler and Keller 2018).

[2] As a more emblematic example of this positive relationship between companies and universities, the author mentions Silicon Valley, in the California region, and the graduates of Stanford University, who changed the relationship between the academic environment and the private sector by demonstrating efficiency in transferring technology to companies, which in turn changed the territorial and economic characteristics of the region.

The concept of shared value recognizes that the needs of the society, not only conventional economic needs, define the market. It concerns the sharing of the value already generated by companies with society (Porter and Kramer 2011).

An innovation is a new or significantly improved product or process, and may also be a combination of them, which differs from the previous products or processes that were made available to potential users or put into use (OECD 2018).

The whole context of changes already underway at the global level due to innovations and technological development signals the unpredictability of new roles and professions at different levels of qualification, while denoting the need for adaptation and rapid response to new demands. For Kon (2020, p. 198), the impact on the labor market affects human resources differently, according to the specific qualification structure of each workforce and according to the internal, regional, and international reaction of each economic space towards changes.

The requalification of roles, relations, and workplaces evoke a series of challenges with regard to the defense of the social, political, and economic interests of workers, which creates changes in labor relations and labor and union legislation to adapt to new situations.

The social perspective demonstrates the potential concerns of companies regarding the desires of the societies in which they operate. In terms of Social Responsibility, it represents the ethical commitment taken on by the company as a procedure for conducting sustainable actions with its stakeholders, such as investments in the community, equal employment opportunities and work safety, among others (Redecker and Trindade 2021; Sion 2021; Husted and Sousa-Filho 2017).

Applying ESG actions, in addition to determining criteria that condition investments, represents a process of cultural transformation in order to understand the fundamental role that companies and organizations can play in society. By adopting ESG practices, a company is valuing issues beyond profit, while also striving for a more sustainable economic performance by valuing environmental, social and corporate governance issues (Redecker and Trindade 2021).

3.8. Questions for reflection and study.

3.8.1.Given the current dynamism, how important is it for professionals to prepare for the future?

3.8.2.What do communities stand to gain with the implementation of the ESG indicators?

3.8.3. In light of the changes and uncertainties regarding the future of work, what should change in academic and professional training for new challenges?

3.8.3. Why is innovation important for organizations and how does innovation make companies stand out in terms of competitiveness?

3.8 Cases to Be Resolved

3.9.1. The Walmart Institute develops social projects to support local and community development programs. Among its projects, the Bombando Cidadania Program has the main goal of raising the Human Development Index (HDI) of the Bairro de Bomba do Hemetério, located on the periphery of Recife (Brazil), from 0.704 to the level of 0.797, which is the average HDIs of the poor communities of the city (Walmart Brasil 2015). Despite the boldness of their goals, what social actions could be undertaken for this purpose?

3.9.2. Currently, one of the aspects to be observed for a company to become competitive and deliver social value is the need for its human resources to provide high levels of productivity and innovative capacity. With the awareness of this need, a certain company hired a consulting team to analyze the skills, abilities, and attitudes of its employees and assess what could be done to generate positive impact in their surrounding community. Based on this scenario, evaluate which strategies could be suggested by the consultancy, based on the ESG perspective.

3.9 Conclusion

The economic transformations that have occurred in recent years have highlighted the importance of innovation in obtaining competitive advantages for organizations, as well as in the strategy for local development and social improvement. In this increasingly dynamic scenario, it is essential for organizations to make their actions more flexible and respond quickly to market demands and society's demands.

Due to the growing inequalities in the capitalist economy, great challenges arise, whether due to changes in modes of production or in the forms of regulation and organization of work. For a long time, we have seen a model that tends to extract value and little internalize environmental and social costs. The transformations in the world of work resulting from innovations and advances in technology have brought to light the precariousness of work, which strains the quality of life in society, as it largely serves the interests of capital. This problem needs to move towards a conception where it is possible to obtain better returns integrated with social progress. The individual, therefore, cannot be considered out of context from their social issues. Not only because work is essential to generate income, but because it must be the means of providing dignity to the way of life.

Social-emotional skills, such as creativity, empathy and the ability to continuously learn, will be increasingly valued as technical skills may become obsolete more quickly. The future of work will be characterized by constant change and adaptation, with new challenges and opportunities emerging. It will be a future that will require flexibility, resilience, and a proactive approach to not succumb to the transformations that are to come.

References

Alves CA, Costa BK (2020) Marketing capabilities and competitive advantage of tourist destinations. Brazilian J Market 19(4):949–984. https://doi.org/10.5585/remark.v19i4.18697

Antunes R (2006) A era da informatização e a época da informalização: riqueza e miséria do trabalho no Brasil. In: Riqueza e miséria do trabalho no Brasil. São Paulo: Boitempo

Axe C, Greve J (2017) Adoção de inovações em contabilidade gerencial: compatibilidade da cultura organizacional e resultados percebidos. Manag Account Res 34(1):59–74

Barbosa A, Costa J, Pontes R (2020) Cidades inteligentes no contexto da quarta revolução industrial. In: A quarta revolução industrial: inovações, desafios e oportunidades. Cadernos Adenauer XXI, 1, Rio de Janeiro: Fundação Konrad Adenauer

Borges MS (2014) Balde Cheio e Proyecto Lechero: Fragilidades e Potencialidades de Políticas Inclusivas e Inovadoras para os Pequenos Produtores de Leite. Tese de doutorado. Universidade Federal Rural do Rio de Janeiro: Seropédica. In: https://tede.ufrrj.br/jspui/bitstream/jspui/2898/2/2014%20-%20Marcio%20Silva%20Borges.pdf

Brogi M, Lagasio V (2018) Environmental, social, and governance and company profitability: are financial intermediaries different? Corp Soc Responsib Environ Manag 26(3):576–587

Caruso L (2018) Digital innovation and the fourth industrial revolution: epochal social changes? AI & Soc 33:379–392. https://doi.org/10.1007/s00146-017-0736-1

Elkington J (2020) Sustentabilidade: canibais com garfo e faca. Books

Engelmann W, Nascimento HCP (2021) The Development of human rights in companies through ESG as a way to qualify labor relations. Revista da Escola Judicial TRT4, Porto Alegre, 3(6):113–135

Firjan (2021) Critérios e métricas ESG para a indústria, Rio de Janeiro, 36p

Frey CB, Osborne MA (2017) The future of employment: how susceptible are jobs to computerisation? Technol Forecast Soc Chang 114:254–280. https://doi.org/10.1016/j.techfore.2016.08.019

Gera I, Singh S (2019) An Inquiry into the impact of the fourth industrial revolution on employment: a review. In: Proceedings of 10th international conference on digital strategies for organizational success. https://doi.org/10.2139/ssrn.3315169

Guerra Filho WS, Turqueti DMC, Lima MVR (2021) A responsabilidade pelo enfrentamento de demandas sociais: a assimilação empresarial do conceito e a nova onda ESG. Revista Jurídica, Curitiba: 1(63):663–683

Half R (2021) Futuro do trabalho: quais as tendências e como preparar a empresa? [Blog post]. https://www.roberthalf.com.br/blog/tendencias/futuro-do-trabalho-conheca-os-s-fatores-que-estao-transformando-o-mundo. Access in February, 2022

Hall BH, Sena V (2017) Mecanismos de apropriabilidade, inovação e produtividade: evidências do Reino Unido. Economia Da Inovação e Novas Tecnologias 26(1):42–62

Harari YN (2018) 21 lições para o século XXI. São Paulo, Companhia das Letras

Hawksworth J, Berriman R, Goel S (2018) Will robots really steal our jobs? An international analysis og the potencial long term impatc of automation. Pricewaterhouse Coopers, London, England, 43p

Henisz W, Koller T, Nuttall R (2019) Five ways that ESG creates value. In The McKinsey Quaterly, Seattle, WA, EUA

Hitt MA, Ireland RD, Hoskinsson RE (2017) Strategic management competitiveness and globalization concepts and cases. 12th Cengage Learning Asia Pte Limited

Hobsbawm E (1995) A Era dos Extremos: O breve século XX – 1914–1991. Companhia das Letras, São Paulo

Husted BW, Sousa-Filho JMD (2017) The impact of sustainability governance, country stakeholder orientation, and country risk on environmental, social, and governance performance. J Clean Prod 155:93–102. https://doi.org/10.1016/j.jclepro.2016.10.025

Kon A (2020) O futuro do mundo do trabalho: impactos do novo paradigma tecnológico. In: A quarta revolução industrial: inovações, desafios e oportunidades. Cadernos Adenauer XXI, 1, Fundação Konrad Adenauer, Rio de Janeiro

Kotler P, Keller KL (2018) Administração de Marketing, 15th edn. Pearson Education do Brasil, São Paulo

Lévy P (1993) *As Tecnologias da Inteligência: o futuro do pensamento na era da informática.* Editora, Rio de Janeiro, 34

Li D, Zheng M, Cao C, Chen X, Ren S, Huang M (2017) O impacto da pressão de legitimidade e lucratividade corporativa na inovação verde: evidências do topo da China 100. J Clean Prod 141(1):41–49

Lueg K, Krastev B, Lueg R (2019) Bidirectional effects between organizational sustainability disclosure and risk. J Clean Prod 229:268–277

Matinaro V, Liu Y (2017) Rumo ao aumento da inovação e sustentabilidade através da cultura organizacional: um estudo de caso de uma empresa de construção finlandesa. J Clean Prod 142(1):3184–3193

Mazon C, Issa RH (2022) Adoção e implementação das práticas ESG (Environmental, Social and Governance) pelas Empresas Estatais: o programa socioambiental da Petrobras e a preservação das comunidades tradicionais. Cadernos Da Escola Paulista De Contas Públicas 1(8):35–52

Natura (2020) Relatório Anual GRI 2020. São Paulo. https://www.natura.com.br/relatorio-anual. Access in February, 2022

Neto JA (2021) A era dos investimentos conscientes: em busca do impacto socioambiental. https://jornal.usp.br/?p=391371. Access in February, 2021

Novakova L (2020) The impact of technology development on the future of the labour market in the Slovak Republic. J Technol Soc 62(101256):1–12. https://doi.org/10.1016/j.techsoc.2020.101256

OECD (2005) Manual de Oslo: Diretrizes para coleta e interpretação de dados sobre inovação. 3 edn. OCDE/FINEP, Brasília

OECD/Eurostat (2018) Oslo Manual 2018: Guidelines for Collecting, Reporting and Using Data on Innovation, 4th edn, The Measurement of Scientific, Technological and Innovation Activities, OECD Publishing, Paris/Eurostat, Luxemburgo. https://doi.org/10.1787/9789264304604-p

Piva M, Vivarelli M (2018) Is innovation destroying jobs? Firm-Level Evidence from the EU. Sustainability 10(4):1279. https://doi.org/10.3390/su10041279

Porter ME (1996) What is strategy? Harvard Bus Rev 74(6):61–78

Porter ME, Kramer MR (2011) Creating shared value. Harv Bus Rev 89(1–2):62–77

Porter ME, Serafeim G, Kramer M (2019) Where ESG fails. In: Institutional investor. In: https://www.institutionalinvestor.com/article/b1hm5ghqtxj9s7/Where-ESG-Fails. Access in December 2021

Redecker AC, Trindade LM (2021) Práticas de ESG em sociedades anônimas de capital aberto: um diálogo entre a função social instituída pela lei n°6.404/76 e a geração de valor. Revista Jurídica Luso-brasileira 7(2)

Schumpeter JA (1982) A Teoria do Desenvolvimento Econômico: uma investigação sobre lucros, capital, crédito, juro e o ciclo econômico. Coleção Os economistas. São Paulo: Editora Abril Cultural

Schumpeter JA (1984) Capitalismo, Socialismo e Democracia. Zahar, Rio de Janeiro

Schwab K (2018) A quarta Revolução Industrial. 1ª ed., editora Edipro. 160p

Shakil MH (2020) Environmental, social and governance performance and stock price volatility: a moderating role of firm size. J Publ, Aff

Sion AO (2021) Compliance ambiental e critérios ESG. In: Sion AO, França, LG (eds) ESG: Novas tendências do direito ambiental. 1ª ed., Synergia, Rio de Janeiro

Sposito ES (2019) Desenvolvimento Regional do Brasil: uma leitura pela ótica da Quarta Revolução Industrial. In: Oliveira FG, et al (Org) Espaço e economia: Geografia Econômica e a Economia Política. Consequência, Rio de Janeiro

Taliente M, Favino C, Netti A (2019) Impacto das Informações Ambientais, Sociais e de Governança no Desempenho Econômico: Evidência de uma 'Vantagem de Sustentabilidade' Corporativa da Europa. Sustentabilidade 11(6):1738. https://doi.org/10.3390/su11061738

The Global Compact (2004) Who cares wins: connecting financial markets to a changing world

Tunes R (2020) Geografia da inovação: território e inovação no Brasil no século XXI. 1. ed. - Rio de Janeiro: Letra Capital: *Observatório das metrópoles*. In: https://www.observatoriodasmetropoles.net.br/wp-content/uploads/2020/03/Geografia-da-Inova%C3%A7%C3%A3o-Territ%C3%B3rio-e-Inova%C3%A7%C3%A3o-no-Brasil-no-s%C3%A9culo-XXI_final.pdf. Access in February 2022

Walmart Brasil (2015) Relatório de Sustentabilidade 2015. In: http://www.mediagroup.com.br/host/walmart/2015/pt/23.htm?s=Bomband%20Cidadania. Access in February, 2022

World Economic Forum (2020, outubro) The future of jobs report. Genebra, Suíça: World Economic Forum. In: https://www.weforum.org/reports/the-future-of-jobs-report-2020. Access in February 2022

Complementary Literature (further reading recommendations)

Arntz M, Gregory T, Zierahn U (2020) Digitalização e o futuro do trabalho: consequências macroeconômicas. In: Zimmermann KF (eds) Handbook of Labor, Human Resources and Population Economics. Springer. https://doi.org/10.1007/978-3-319-57365-6_11-1

Cornell B, Damodaran A (2020) Valuing ESG: doing good or sounding good? J Impact e ESG Investing Fall 1(1):76–93. https://doi.org/10.3905/jesg.2020.1.1.076

Corrêa MM, Ashley PA (2018) Desenvolvimento Sustentável, Sustentabilidade, Educação Ambiental e Educação para o Desenvolvimento Sustentável: Reflexões para ensino de graduação. REMEA - Revista Eletrônica Do Mestrado Em Educação Ambiental 35(1), 92–111. https://doi.org/10.14295/remea.v35i1.7417

Da Costa T, Alperstedt GD, Andion C (2021) Dimensions of the approach: a conceptual proposal. Revista Brasileira De Gestão e Desenvolvimento Regional 17(1):165–178

Garcia A, Orsato RJ, Lugoboni L (2018) O Desempenho Empresarial nos fatores "ESG-Environmental, Social and Governance" em diferentes ambientes institucionais. In: XLII Encontro da ANPAD - EnANPAD. Editora. ANPAD

Glinskiy VV, Serga LK, Alekseev MA (2020) Territorial differentiation as the factor of sustainable economic development. Procedia Manuf 43:263–268. https://doi.org/10.1016/j.promfg.2020.02.152

Hill J (2020) Environmental, social, and governance (ESG) investing: a balanced analysis of the theory and practice of a sustainable portfolio. Academic Press

Hunt W, Sarkar S, Warhurst C (2022) Measuring the impact of AI on jobs at the organization level: Lessons from a survey of UK business leaders. Res Policy 51(2):104425. https://doi.org/10.1016/j.respol.2021.104425

Janssen S, Mohrenweiser J (2018) The shelf life of incumbent workers during accelerating technological change: evidence from a training regulation reform. IZA Discussion Paper, 1312. https://doi.org/10.2139/ssrn.3129256

Joseph GPA, Gonçalves-Dias SLF, Felsberg AV, Igari AT (2018) Responsabilidade Social Corporativa e índices de sustentabilidade: um estudo dos ativos tangíveis e intangíveis à luz da visão baseada em recursos. In: Revista de Gestão Social e Ambiental – RGSA, São Paulo 12(1):73–88. https://doi.org/10.24857/rgsa.v12i1.1304

Kahupi I, Hull CE, Okorie O, Millette S (2021) Building competitive advantage with sustainable products—a case study perspective of stakeholders. J Cleaner Product 289(125699). https://doi.org/10.1016/j.jclepro.2020.125699

McManus J (2021) How do innovation, technology, and competitiveness contribute to business growth. In: Ferreira JJM, Teixeira SJ, Rammal HG (eds) Technological Innovation and international competitiveness for business growth. https://doi.org/10.1007/978-3-030-51995-7_3

Michelle Do Carmo Vieira Master's student in the Territorial Development and Public Policy Post-graduate program of the Universidade Federal Rural do Rio de Janeiro (PPGDT/UFRRJ). She is an Administration graduate by the Universidade Federal Rural do Rio de Janeiro (UFRRJ) and has a full teaching degree in Geography from the Universidade do Estado do Rio de Janeiro (UERJ). She has specialized in the Teaching of Geography in the Universidade do Estado do Rio de Janeiro (UERJ) and acts as a Basic Education teacher in the private sector.

Marcio Silva Borges Post-Doctor in Company Management and Economic Management of Natural Resources by the Universidad de Lleida (UdL/Spain), a Doctor in Technology and Innovation Sciences in Agribusiness by the Universidade Federal Rural do Rio de Janeiro (UFRRJ), graduate in Administration by the Universidade Federal Fluminense (UFF), graduate in Rural Management and Development by the Universidade Federal do Rio Grande do Sul (UFRGS). He acts as an Associate Faculty member in the Department of Tourism and Administration (DAT/UFRRJ) and a permanent faculty member in the Post-Graduate Program for Administration and Tourism, as well as in the post-Graduate program for Territorial Development and Public Policy (PPGDT/UFRRJ). It's a member of the Red Internacional en Tecnología, Innovación, Competitividad y Sostenibilidad (RedITICS).

Part III
Social Perspective

Chapter 4
Environmental, Social, and Governance (ESG) Risks Arising from Contingent Liabilities Related to Environmental Disaster in Companies: Mining Cases in Brazil

Sergio Ricardo Da Silveira Barros and Sergio Luiz Braga França

Abstract The text emphasizes the need for organizations to adopt policies for environmental management, safety, and occupational health as part of their social responsibility. It also discusses the broader context of environmental risks in the modern world and their implications for organizations and society at large. Additionally, it highlights the importance of transparency, stakeholder engagement, and proactive risk management in addressing environmental challenges and conflicts. The objective of the work is to relate ESG strategies to Environmental, Social and Governance Risks (ESG) arising from contingent liabilities related to environmental disasters in companies: mining cases in Brazil. In summary, the text underscores the need for a more comprehensive and balanced approach to ESG, where organizations not only focus on the positive aspects of sustainability but also account for and actively address the negative impacts of their operations. By prioritizing social responsibility, engaging communities, and promoting a healthy environment, organizations can work toward environmental sustainability while reducing the financial risks associated with potential damages and conflicts.

Keywords ESG · Environmental conflict · Contingent liabilities

S. R. Da Silveira Barros (✉)
Fluminense Federal University, Niterói, RJ, Brazil
e-mail: sergiobarros@id.uff.br

S. L. B. França
Civil Engineering, Fluminense Federal University, Niterói, RJ, Brazil
e-mail: sfranca@id.uff.br

M. J. Meiriño and O. L. G. Quelhas (eds.), *Critical Factors Associated with Environmental, Social and Governance Management in Organizations*, Accounting, Finance, Sustainability, Governance & Fraud: Theory and Application, https://doi.org/10.1007/978-981-96-1725-8_4

4.1 Role of the Chapter in the Structure of the Book

The new demands of economic agents have been suggesting that organizations adopt policies for environmental management, safety, and occupational health, thus seeking to create more sustainable scenarios within the framework of Social Responsibility. To this end, policies that integrate business management with territory management are necessary. One of the paths that have been proposed is participatory management through stakeholders in issues that affect them directly, such as environmental risks (Barros et al. 2010). Currently, environmental disasters from mining activities in Brazil have drawn attention due to the extent of damage and environmental conflicts in communities that need the natural resources of the areas of influence of these projects and the treatment of their risk management policies (Pinto 2019).

It is observed when investigating these events, the focus is fixed on the moment of occurrence of the event, without taking into account the procedures previously adopted and the proactive management actions, often neglected and that led to the occurrence of the accident according to Duarte (2002), ethical and psychological aspects must also be investigated after an accident, unfortunately, it is through the accident that many organizations change their attitude. A high severity event generates significant learning in the work environment that, due to their complexity, these expanded events have common characteristics: fatal victims or not in the workforce; fatalities or not, among the members of the local communities and damage to the environment in great extensions. These events generate in the organizations' financial statements the need to create a contingent liability due to the need to provision for the damages.

4.2 Risk Society

Organizations today face unprecedented risks as stakeholders demand accountability and transparency in the way companies approach the environment, care for the well-being of their workers and the community; and whether they govern their affairs in an ethical manner.

The risk of modern society is something non-tangible, existing for everyone, regardless of social class, however, privileged groups adopt strategies against threats, being able to afford to choose where to live, the food they will consume and the water they will drink. The strategies of the privileged classes do not guarantee the suppression of risk, not least because some global threats may affect everyone indiscriminately. In this case, one can mention the risk to which hydrographic basins all over the planet will be exposed, with the possibility, in a few years, of the loss of water quality for different uses.

Risk, according to Beck (1992) and Fabiani (2003), is fully incorporated today's society, which is considered a risk society. For him, risk is the main axis of social organization and knowledge (information), the main axis of risk. This theory, by the

German sociologist, considers that society puts itself in the hands of expert systems, by Giddens (2002), which do not present a solution to the environmental crisis, but force information and risk assessment for those in power.

Another important aspect of Beck's Risk Society Theory is that the new techniques generated the rupture of the four pillars of private insurance: compensation, anticipation of consequences, temporal and spatial limitation and strict causality with the individualization of responsibilities. Another important aspect of Beck's Risk Society Theory is that the new techniques generated the rupture of the four pillars of private insurance: compensation, anticipation of consequences, temporal and spatial limitation and strict causality with the individualization of responsibilities. Major environmental disasters often reveal the fragility of control structures and the normalized increase in global threats.

How to define environmental risk? What factors to consider for your analysis? There is no way to consider environmental risk only as that which affects the internal environments of the organization, as it is not the risk that affects the spaces around the company, nor is it that which may affect the health of workers and communities. Environmental risk is a symbolic category, built according to each local reality, which arises through the interactions of all risks related to an organization and its environment, considering the various social, economic, ecological, cultural, and political dimensions, which are given in the middle. Environmental risks and conflicts occur due to the appropriation of territories, by economic activities whose impact generates externalities for society, with negative externalities causing the degradation of the local environment.

According to Egler (1996), Castro et al. (2005), the analysis of environmental risk is a dynamic indicator, which has relationships between natural systems, production systems, and social conditions in each place and time. The elaboration of a comprehensive conception it should consider the vulnerabilities of natural systems, the relationship between economic parameters and the use of territories by organizations and the systematic duality between human development and social issues.

Environmental sustainability, according to Sulaiman and Aledo (2016), will occur when the significant importance of social commitment is officially recognized, where environmental standards are previously agreed through participatory processes involving many people. In this way, the risks occurring in the coastal zone can only be evaluated if they consider the different rationality matrices, Gonçalves (2002, 2004) in their risk management plans, given the diversity of social actors currently in this space.

To manage the risks associated with ESG (Environmental, Social and Governance), three issues are important to assess. What environmental, social and governance issues are financially material? How are companies dealing with these material risks? How will these risks affect the long-term value of companies? An important aspect of governance, in the context of ESG, is the recognition of contingent liabilities, which, according to De Souza (2017), consist of an obligation resulting from past events for which the organization is responsible, and which must be disclosed in explanatory notes in the accounting balances due to the uncertainties of the amounts attributed to the contingency.

Based on the importance of ESG in organizations, this work aims to discuss contingent liabilities and environmental conflicts caused by environmental disasters in mining activities in Brazil.

4.3 Concepts on Environmental Impacts, Damages, and Conflicts

According to Hannigan (2009) contemporary society has realized the damage caused by its economic model disconnected from environmental and social issues, causing grave consequences for maintaining the balance of planetary ecosystems. The author established a model that became known as Competitive Functions of the Environment.

Hannigan (2009) proposes is that contemporary society advances on spaces without realizing their finitude, throwing residues from all its activities into the environment, while using natural resources in large proportions for their activities, economic processes. This model is what in economic science generates scarcity, that is, a good whose demand far exceeds supply. It can be said that scarcity is one of the great motivators of contemporary socio-environmental conflicts, especially when natural resources are extracted and traded without any, or little concern, with people and the environments in which they live.

In the process of its reproduction, societies confronted with different projects for the use and significance of their resources, which means admitting that the environmental issue is intrinsically conflicting, although this is not recognized (Viégas 2009). Changes in the environment only become environmental damage in social processes in which collective subjects define them as such. (*op cit*). It is important to clarify that there is a sequential relationship between economic activity, its environmental impacts, its damage and, consequently, its generated conflicts (Noranha et al. 2018; França and Barros 2018) (Fig. 4.1).

The model above suggests that all economic activity causes environmental impacts, that is, changes in the environment, both in the extraction of natural resources and in the final disposal of the productive processes of what was not used in a product or service. When this final disposal occurs without due care, they end up generating, in many cases, environmental damage, which can be identified as the materiality of the environmental impact. This materiality will be due to changes in the physical, chemical, and biological properties of the environment, resulting from human activities that, directly or indirectly, affect the health, safety, and well-being of the population; social and economic activities; the natural environment; the aesthetic and sanitary conditions of the environment and the quality of environmental resources.

In this context, environmental conflicts arise, given the different ways of using the spaces where these activities are carried out, with political-economic interest being the main vector. According to Turner (2004), what is considered fundamental for the understanding and configuration of the trajectory of environmental conflicts that are

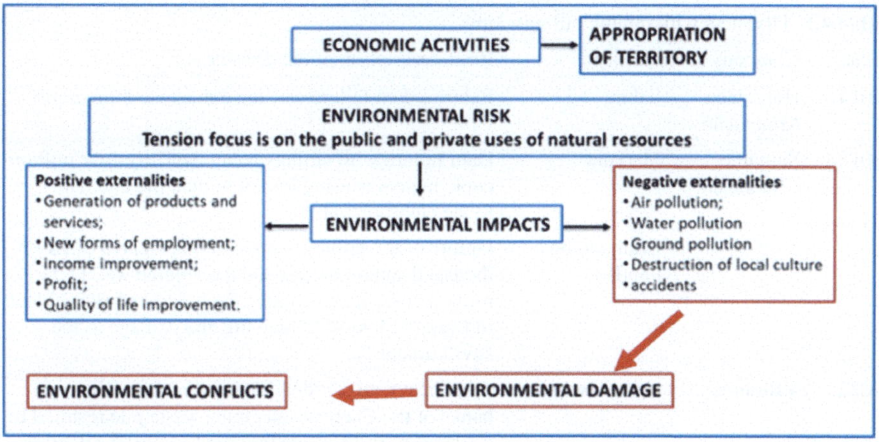

Fig. 4.1 Model for the formation of environmental conflicts. *Source* Noronha et al. (2018)

analyzed: the interest groups involved (social actors), their strategies, powers, and vulnerabilities.

According to Gomes (2002), the magnitude of the appropriation of a space can range from a simple occupation of an area to the domain of a territory occupied by traditional populations. And the manifestation of the appropriation of the territory can occur through fixed structures, as well as through an appropriation conveyed by more subtle and symbolic instruments.

4.4 History of the Development of the Theme

Currently, according to the Environmental Justice Atlas (2022), Brazil has mapped 174 environmental conflicts arising from various economic activities, including mining. After the collapse of the Mariana and Brumadinho tailings dam in 2019, a group of researchers and activists from various parts of the world collectively produced a thematic map showing thirty cases of environmental conflicts in which the main mining company in Brazil has or had some responsibility (SAES et al. 2021). Allied to the concept of environmental conflict is the Environmental Justice, which describes the disproportionate power of environmental risks to populations that have lower incomes, little political power, and lack of information (ACSELRAD 2011).

Santos and Oliveira (2021) warn that the larger is the tailing dam, as greater is the concern about its geotechnical management, since failure of these structures can lead to catastrophic consequences including human, environmental, and economic loses. The work of Santos and Wanderley (2016) state that from the "commodities boom" in 2011, it constituted the main element inducing the expansion of infrastructure

Table 4.1 Disasters with mining tailings dams

Year	Company	locality	Brief description of the disaster
2014	Herculano Mineração	Itabira	Itabirito dam failure causing damage to human life
2015	Samarco Mineração	Mariana	Dam failure with tailings volume of fifty-four million cubic meters causing loss of human life and damage to the environment
2019	Vale	Brumadinho e Fundão	Failure of the tailings dam stored in an area of 249.5 thousand square meters and the volume was 11.7 million cubic meters, being the biggest work accident in Brazil in loss of human life and damage to the environment
2022	Vallourec	Nova Lima	After heavy rain, a slope failure involving three banks of the Cachoeirinha mine waste pile resulted in a release of mine waste into the Lisa water retention dam. The dam then overflowed on the whole length of the embankment, without compromising its stability. There was no failure in the tailings dam, but the mud wave that overflowed the retention dam blocked the BR-040 highway, which was closed for almost two days

Fonte: based on Santos e Wanderley (2016); Agência Brasil January 25, 2022 in https://agenciabrasil.ebc.com.br/geral/noticia/2022-01/chuvas-colocam-barragens-em-alerta-3-anos-apos-ruptura-em-brumadinho e Thompson et al (2020)

associated with mining, with the inherent risk of the operation, the allocation of tailings in dams. Next, Santos & Wanderley list the accidents resulting from ruptures of tailings dams in the state of Minas Gerais, Brazil, up to the year 2015, complemented by this work with the accident in Brumadinho, also in the state of Minas Gerais in January 2019. Below, the most impactful disasters involving mining tailings dams in the state of Minas Gerais will be presented (Table 4.1).

The recognized lofty standards of Corporate Social Responsibility of large Brazilian companies in the mining sector coexist with several environmental conflicts that generate claims from vulnerable and indigenous populations, whose lives and livelihoods are threatened by environmental degradation and the expansion of mineral extraction (SAES et al. 2021).

4.5 Environmental Risk and Contingent Liabilities

Accounting disclosure should involve the context of companies in disclosing not only the reports required by legislation, but also tables and voluntary information, thus illustrating the concern to be as transparent as possible to society, that is, to legitimize itself (Deegan and Rankin 1996). Thus, contingent liabilities must be clearly demonstrated to show the provisioning of liabilities that organizations will have to bear in the future.

It should be noted that Lemos et al. (2019) evaluated that despite the related evidence, the lack of changes in topics on risk management of mining activity and loss of socio-environmental information as an integral part of the company's financial statements as well as the Social Balance. It also verified that relevant changes to the risks related to socio-environmental issues are released only in the reference forms (contingent liabilities), which is a document that is not as comprehensive as the standardized financial statements, and that do not demonstrate the risks to the population close to the dams, nor that their failure can cause damage to the environment and human life.

The disclosure of environmental and social information by organizations has become a growing and voluntary practice. Even before the emergence of the environmental issue, the academy was already directing its studies to this theme. In this way, investors are increasingly looking for ESG criteria in companies' portfolios so that their decisions are guided by business ethics and that promote adequate effort without sacrificing the return on investment (Pedersen et al. 2021). It should be noted that ESG information disclosures are still lacking in reliability, comparability, and relevance from the perspectives of investors. Many ESG practices primarily align with corporate social responsibility (CSR) measures leading to better organizational performance and improved quality of accounting information.

Due to the repercussion of environmental damage, most investors began to direct their resources to those socially responsible companies, whose published accounting reports demonstrate that measures have been taken to reduce the impacts that economic growth causes to the natural and social environment (Gomes et al. 2019).

The ESG risks inherent to the provisions for contingent liabilities can be significant and, therefore, must be recognized and considered in the analyzes and properly disclosed to all stakeholders, to minimize the impacts arising from this information that is outside the balance sheet and therefore not covered in the income statement for the period.

4.6 Methodology for Researching Cases of Accidents and Conflicts

As for the procedures, the documentary and bibliographic analysis was used as a way of investigating the documents to understand the phenomenon. To achieve the objective of this work, research within the literature, empirical research, and exploratory analysis of the collected material were conducted. Figure 4.2 schematically presents the flow of the adopted research method.

A search was conducted in the SCOPUS database, associating "environmental conflict" and "mining" in the subarea of Social Sciences, and 149 articles were found. When the word "Brazil" is added to the Boolean search, nineteen documents are found, after being analyzed, eleven articles were selected that have a direct relationship between mining activity and socio-environmental conflicts and, consequently, environmental damage is already materialized.

Fig. 4.2 Methodology steps. *Source* Authors

4.7 Discussion on the Consequences of Environmental Disasters and Conflicts Generated by Mining

The following articles were listed as a way of discussing the impacts, damages and conflicts arising from mining activities in Brazil, evaluating their effects on society and the environment. The following articles were allocated in chronological order of publication (Box 4.1).

Box 4.1: Selected cases on the consequences of environmental disasters and conflicts generated by mining

1	Chizenga, A.P., Blanco, G., Almeida, J **The category of place and its relevance to social sciences: reflections on environmental conflicts in Moatize (Mozambique) and Araxá (Brazil)** [Article@A categoria de lugar e sua relevância para as ciências sociais: uma reflexão a partir de conflitos ambientais em Moatize (Moçambique) e Araxá (Brasil)] (2022) Horizontes Antropologicos, 28 (63), pp. 359–396.1) https://www.scopus.com/inward/record.uri?eid=2-s2.0-85132225249&doi=10.1590%2fS0104-71832022000200013&partnerID=40
2	Ribeiro, A.P., Ruiz, M.S., Cortese, T.T.P., da Silva, L.F., Aranha, S., Zei, M.L.F **Mediation of socio-environmental conflicts in Paraná: Mining and impacts on public Health** [Article@Mediação de conflitos socioambientais no Paraná: Mineração e impactos à saúde pública] (2021) Urbe, 13, art. no. e20200318,. Cited 1 time https://www.scopus.com/inward/record.uri?eid=2-s2.0-85117892378&doi=10.1590%2f2175-3369.013.e20200318&partnerID=40
3	Marques, N., Fazito, M., Cunha, A **Tourism development discourse dynamics in a context of conflicts between mining and nature conservation in the Brazilian Cerrado Hotspot** (2021) Journal of Sustainable Tourism,. Cited 2 times https://www.scopus.com/inward/record.uri?eid=2-s2.0-85106329546&doi=10.1080%2f09669582.2021.1914066&partnerID=40&
4	De Los Guevara, M.A., Moreira, E.S **(Neo)extractivism and its impact on the lives of women in the southeast of Pará State (Brazil)** [Article@O (neo)extrativismo e seu impacto nas vidas das mulheres no sudeste do Pará] (2020) Desenvolvimento e Meio Ambiente, 54, pp. 227–248. Cited 1 time https://www.scopus.com/inward/record.uri?eid=2-s2.0-85095454465&doi=10.5380%2fDMA.V54I0.66101&partnerID=40&md5=1ac36c7ea3a9247c0601383d5ea055ef

(continued)

(continued)

5	De Macêdo Coelho, Y.C., Lucas, F.C.A., De Medeiros Sarmento, P.S **Environmental perception and mining of aggregates: The view of the urban–rural population of Ourém, Pará, Brazil** [Article@Percepção ambiental e mineração de agregados: O olhar da população urbano-rural de Ourém, Pará, Brasil] (2020) Desenvolvimento e Meio Ambiente, 53, pp. 38–60. Cited 1 time https://www.scopus.com/inward/record.uri?eid=2-s2.0-85088387616&doi=10.5380%2fdma.v53i0.60771&partnerID=40&md5=d952722c01422ec8ca863b7839f60ff0
6	Urquiza, A.H.A., de Oliveira Rocha, A **The environmental disaster of Mariana and the krenaks of the Doce River** [Article@O desastre ambiental de mariana e os krenak do rio doce] (2019) Veredas do Direito, 16 (35), pp. 191–218 https://www.scopus.com/inward/record.uri?eid=2-s2.0-85077561549&doi=10.18623%2frvd.v16i35.1507&partnerID=40&md5=7437dd615ba3dd9ab6b3da1e514c0cf7
7	Rojas, C.M.O., Pereira, D.B., Dias, J.B **The conflicts of (Neo)extractivism: Mines, tailings dams, and protected areas in the metropolitan area of Belo Horizonte, Brazil** (2019) CyberGeo, 2019, https://www.scopus.com/inward/record.uri?eid=2-s2.0-85075471818&doi=10.4000%2fcybergeo.32684&partnerID=40&md5=779796
8	da Rocha, D.F., Porto, M.F., Pacheco, T., Leroy, J.P **The map of conflicts related to environmental injustice and health in Brazil** (2018) Sustainability Science, 13 (3), pp. 709–719. Cited 7 times https://www.scopus.com/inward/record.uri?eid=2-s2.0-85031497830&doi=10.1007%2fs11625-017-0494-5&partnerID=40&md5 =
9	Costa, D.M **Masquerade as memory: Conflict as heritage in Lavras do Abade, Brazil** (2016) International Journal of Intangible Heritage, 11, pp. 116–126 https://www.scopus.com/inward/record.uri?eid=2-s2.0-84994893902&partnerID=40&md5=f715c3823988d4762ea01f05ba18ca3b
10	Almada, E.D., Anaya, F.C., Monteiro, F.T **The people of the mountains: The biocultural heritage of the Espinhaço range in Minas Gerais state, Brazil** (2016) Ecology and Conservation of Mountaintop Grasslands in Brazil, pp. 479–500. Cited 4 times https://www.scopus.com/inward/record.uri?eid=2-s2.0-84978245691&doi=10.1007%2f978-3-319-29808-5_20&partnerID=40&md5
11	Bacci, D.C., Diniz, T.M.R.G Mining in urban areas: **Methodological proposal for the identification and mediation of socio-environmental conflicts** [Article@Mineração em áreas urbanas: Proposta metodológica para identificação e mediação de conflitos socioambientais] (2013) Revista Escola de Minas, 66 (3), pp. 369–374. Cited 5 times https://www.scopus.com/inward/record.uri?eid=2-s2.0-84881486809&doi=10.1590%2fS0370-44672013000300015&partnerID=40

It is important to emphasize that the articles list the environmental conflicts arising from mining activity in Brazil, bringing to this work the great question about the effective importance of ESG practices linked to risk management and the effective measurement of contingent liabilities, especially after the great disasters that occurred in the state of Minas Gerais with the failure of mining tailings dams in Mariana and Brumadinho.

According to Ribeiro et al (2021) reveal that "The socio-environmental conflict in SMS, among others observed in Brazil, reveals the distance from what has been strongly proclaimed as an ethical, social, responsible, and environmentally conscious corporate behavior of mining companies, and what is happening to people and the environment in contexts of subalternity. Actors directly affected by exploration activities, such as traditional populations, small farmers, residents, among others, have been historically and socially ignored from the social and business responsibility strategies indicated by large developers. The authors portray that the decision-making difficulties in relation to the search for solutions to socio-environmental conflicts, in part, reflect the lack of systematic use of environmental planning in the phase preceding the environmental licensing of projects, a management stage in which consultation, clarification debates and the composition of interests can be incorporated naturally, due to the availability of time necessary for this type of enterprise," which demonstrates the great asymmetry between the enterprise and the affected population since the conflicts were not resolved.

De Los Guevara and Moreira (2020) "demonstrate the potential of women in strategies supported by solidarity, which leads them to reflect as political subjects. The diversity of life trajectories marked by the spatial movement is no obstacle to articulating women's collective actions, expressions of their agencies in the face of the capitalist and patriarchal hegemonic model that oppresses them. The groups studied constitute spaces for democratic participation, solidarity, autonomy, and expression of women's actions in the face of socio-environmental conflicts and violence that affect them, thus reconstructing other ways of being and relating to nature." Once again, the asymmetries between the economic and the social, especially in populations that use natural resources as a source of their existence.

According to De Macêdo Coelho et al (2020) "social actors mentioned more damages than benefits from mining in the municipality, which is evidenced by the derisory measures to repair damages, low employability in enterprises and lack of actions to improve quality. of life of the population. The damages of the activity to the physical environment are the most noticed, due to the changes in the landscape and in the ways of life; however, the population does not effectively feel the adverse effects of the health sector. Popular articulation to demand reparation/mitigation of damages and social counterparts is rare and punctual. People have shown a tendency toward activist and/or questioning attitudes only when they are most directly affected." The model of asymmetry between the economic, the social and the environmental is always repeated, but when it comes to conflicts, the social has the greatest damage.

According to Urquiza et al. (2019) specifically with regard to the Mariana disaster, it was found that, when dealing with the tragedy, the government and the companies that caused the damage made no distinction regarding the indigenous peoples, including acknowledging the peculiarities of the relationship between the Doce River and the Krenak population, whose physical and cultural life takes place on its margins, in addition to other ethnicities, such as the Tupiniquim, who live in the neighboring state of Espírito Santo, and were also affected by the tragedy. In the case of the affected Indigenous peoples, especially the Krenak, the responsibility of the mining companies goes beyond the mere reparation aspect, to become a case of a need for restoration of violated Human Rights, because the contamination and the unfeasible condition of access to Doce River, or "Uatú," compromise the very physical and spiritual existence of that population. Once again, there is a lack of governance of the mining company with the native peoples, even after the disaster that environmentally disabled the Rio Doce for the community.

According to Rojas et al. (2019) "As we have demonstrated, under the argument of "public interest," (neo) extractive-based development policies deprive local communities of the access, use, and control of natural resources and their territory. The rhetoric of "public interest," used in developmental discourse, has been applied to legitimize the social reproduction of models inspired by certain groups, classes, and countries. However, "public interest" is nothing more than the interest of a specific group that presents its demands as universal. In addition, such a policy departs from the separation of man/nature, excludes local communities, and threatens their survival. It clearly shows that there is an unequal distribution of power in terms of access to and use of natural resources and territory, thus promoting environmental injustices and favoring the emergence and proliferation of socio-environmental conflicts."

According to Bacci et al. (2013), "One of the aspects that increase conflicts in urban areas where mining is predominant is when government and society have an immediate and fragmented vision, which seek to "solve a problem," without considering that other problems may arise. A complex view is necessary when environmental issues are considered. Also, partial, and short-term solutions postpone or extend the conflicts. The results point to a continuous process, because as actors perceive more, they participate more and acquire more possibilities of negotiation. Since a shared understanding of the complexity of the process is that social actors may find ways to solve or minimize the environmental conflicts. For a socio-environmental sustainability, solutions must be collaborative and participatory. This takes time and willing ness to dialogue and negotiate." The article provides an overview of the importance of negotiating conflicts between actors, which should be the way to improve governance in the territory.

The Observatory of Mining Conflicts in Latin America—an organization made up of forty organizations whose objective is to defend communities affected by mining—states that more than 211 conflicts caused by mining in 2018 are registered throughout the continent, companies have lost their social licenses for the works. and permission from communities. This rejection is happening due to an awareness in the communities that mining is an activity that has harmful consequences for the population.

4.8 Reflections on Environmental Conflicts and Contingent Liabilities

The articles bring an asymmetry of the ESG in the management of damages and conflicts of the organizations with the territory to which they are inserted, it should not be forgotten that economic activities generate positive and negative externalities in their areas of influence; therefore, a review of provisions for damages (contingent liabilities) in its financial statements as a form of guarantee for compensation for damages caused by its activity.

The social issue should be relevant in ESG models, a healthy environment and social equity will give organizations an economic return that will minimize risks, without the need to reserve amounts for contingent liabilities in their income statements. Environmental sustainability will occur when the entrepreneur recognizes the great importance of the commitment to its social responsibility policy, where through participatory processes communities participate more actively in territorial management programs.

References

Acselrad H (2011) O que é justiça ambiental. 160 p. Garamond. Rio de Janeiro, RJ – Brasil – ISBN 857617597

Almada ED, Anaya FC, Monteiro FT (2016) The people of the mountains: the biocultural heritage of the Espinhaço range in Minas Gerais state, Brazil. In: Ecology and conservation of mountaintop grasslands in Brazil. Springer, Cham. pp 479–499 https://doi.org/10.1007/978-3-319-29808-5_20

Atlas EJ (2022) Global Atlas of environmental justice. https://ejatlas.org/

Bacci DDLC, Diniz TMRDG (2013) Mining in urban areas: methodological proposal for the identification and mediation of socio-environmental conflicts. Rem: Revista Escola de Minas 66:369–374. https://doi.org/10.1590/S0370-44672013000300015

Barros SR, Wasserman JCF, Lima GBA (2010) Environmental risk in the coastal zone: a participating management interdisciplinary proposal for the plans for control of emergencies in organized Brazilian harbours. J Integrated Coastal Zone Manage 10(2):217–227. https://doi.org/10.5894/rgci173

Beck U (1992) From industrial society to the risk society: questions of survival, social structure, and ecological enlightenment. Theory Cult Soc 9(1):97–123

Castro CM, de Oliveira Peixoto MN, do Rio GAP (2005) Environmental risks and geography: conceptualizations, approaches and scales. Anuário do Instituto de Geociências, 28(2):11–30. ISSN 0101-9759

Chizenga AP, Blanco G, Almeida J (2022) The category of place and its relevance to social sciences: reflections on environmental conflicts in Moatize (Mozambique) and Araxá (Brazil). Horizontes Antropológicos 28:359–395. https://doi.org/10.1590/S0104-71832022000200013

Costa DM (2016) Masquerade as memory: conflict as heritage in Lavras do Abade. Brazil Int J Intangible Heritage, 11:116–126. https://www.scopus.com/inward/record.uri?eid=2-s2.0-84994893902&partnerID=40&md5=f715c3823988d4762ea01f05ba18ca3b

da Rocha DF, Porto MF, Pacheco T, Leroy JP (2018) The map of conflicts related to environmental injustice and health in Brazil. Sustain Sci 13(3):709–719. https://doi.org/10.1007/s11625-017-0494-5

De Los Guevara MA, Moreira ES (2020) (Neo)extractivism and its impact on the lives of women in the southeast of Pará State (Brazil) O (neo)extrativismo e seu impacto nas vidas das mulheres no sudeste do Pará] Desenvolvimento e Meio Ambiente 54:227–248

De Souza Ribeiro M, Junior MDST (2017) Os efeitos econômicos do rompimento de barragem de resíduos: divulgações nas demonstrações contábeis comparativamente à grande mídia. Revista De Administração Da Universidade Federal De Santa Maria 10:100–116. https://doi.org/10.5902/1983465925338

de Macêdo Coelho YC, Lucas FCA, de Medeiros Sarmento PS (2020) Percepção ambiental e mineração de agregados: o olhar da população urbano-rural de Ourém, Pará, Brasil. Desenvolvimento e Meio Ambiente 53:38–60. https://doi.org/10.5380/dma.v53i0.60771

Deegan C, Rankin M (1996) Do Australian companies report environmental news objectively? An analysis of environmental disclosures by firms prosecuted successfully by the Environmental Protection Authority. Account Audit Accountab J. https://doi.org/10.1108/09513579610116358

Duarte M (2002) Riscos industriais: etapas para a investigação e a prevenção de acidentes. FUNENSEG., Rio de Janeiro. ISBN 857052384X

EGLER CAG (1996) Risco Ambiental como Critério de Gestão do Território. Revista Território 1: 31–41

Fabiani JL (2003) La société du risque. Sur la voie d'une autre modernité: U. Beck (préface de B. Latour) Paris, Aubier, 2001, 522 p. et. https://doi.org/10.4000/questionsdecommunication.7281

França RF, Barros SR (2018) Gerenciamento de conflitos para a gestão de praias: um estudo de caso na praia de Camboinhas, Niterói, RJ. Desenvolvimento e Meio Ambiente, 44:325–339. https://dx.doi.org/10.5380/dma.v44i0.54923

Giddens A (2002) As consequências da modernidade. São Paulo: 1ª ed. UNESP, 2002. ISBN 85-7139-022-3

Gomes Paulo César da Costa (2002) A condição urbana: ensaios de geopolítica da cidade. Bertrand Brasil, Rio de Janeiro. ISBN-10 8528609561

Gonçalves CWP (2002) Os (des) caminhos do meio ambiente. Editora contexto. 10ªed ISBN-10 8585134402

Gonçalves CWP (2004) O Desafio ambiental. Editora Record; 6ª edição ISBN-13 978-8501069412

Gonçalves CWP (2010) Os (des) caminhos do meio ambiente. Editora contexto. 10ªed ISBN-10 8585134402

Hanningan J (2009) Sociologia Ambiental. Editora Vozes: Rio de Janeiro. 272p.ISBN-13 978-8532638069

Lemos VDS, Klug YS, Da Cruz APC, Barbosa MAG (2019) Conservadorismo contábil na legitimação das empresas de capital aberto do setor de minério no Brasil. Contabilidade Vista & Revista 30(2):113–142. https://doi.org/10.22561/cvr.v30i2.4842

Marques N, Fazito M, Cunha A (2021) Tourism development discourse dynamics in a context of conflicts between mining and nature conservation in the Brazilian Cerrado Hotspot. J Sustainable Tourism, 1–23. https://doi.org/10.1080/09669582.2021.1914066

Noronha G, Barros SRS, Santos RF (2018) Uma avaliação dos conflitos socioambientais no Núcleo Massambaba do Parque Estadual da Costa do Sol. Desenvolvimento e Meio Ambiente, 44. https://dx.doi.org/10.5380/dma.v44i0.54922

Pedersen LH, Fitzgibbons S, Pomorski L (2021) Responsible investing: The ESG-efficient frontier. J Financ Econ 142(2):572–597. https://doi.org/10.1016/j.jfineco.2020.11.001

Pinto RG (2019) Conflitos ambientais, corporações e as políticas do risco. 458p. GARAMOND: Rio de Janeiro, RJ, Brasil. ISBN 9788576174813

Ribeiro AP, Ruiz MS, Cortese TTP, Silva LFD, Aranha S, Zei ML. F. (2021) Mediation of socio-environmental conflicts in Paraná: mining and impacts on public health. Urbe. Revista Brasileira de Gestão Urbana, 13. https://doi.org/10.1590/2175-3369.013.e20200318

Rojas Orduz CM, Barros Pereira D, Bruno Dias J (2019) The conflicts of (Neo) extractivism: Mines, tailings dams, and protected areas in the metropolitan area of Belo Horizonte, Brazil. Cybergeo: Eur J Geography. https://doi.org/10.4000/cybergeo.32684

Saes BM, Bene DD, Neyra R, Wagner L, Martínez-Alier J (2021) Justiça ambiental e irresponsabilidade social corporativa: o caso da mineradora Vale SA. Ambiente & Sociedade 24. https://doi.org/10.1590/1809-4422asoc20210014vu2021L4ID

Santos RSP, Wanderley LJ (2016) Dependência de barragem, alternativas tecnológicas e a inação do estado: repercussões sobre o monitoramento de barragens e o licenciamento do Fundão Rodrigo. Editorial iGuana. ISBN 978 85 68819 09 8

Santos TB, Oliveira RM (2021) Failure risk of brazilian tailings dams: a data mining approach ENGINEERING SCIENCES. An Acad Bras Ciênc 93(4):2021 https://doi.org/10.1590/0001-3765202120201242

Santos RF, da Silveira Barros SR (2018) Gerenciamento de conflitos para a gestão de praias: um estudo de caso na praia de Camboinhas, Niterói, RJ. Desenvolvimento e Meio Ambiente 44:325–339. https://doi.org/10.5380/dma.v44i0.54923

Santos GN, Barros SRS, Santos RF (2018) Uma avaliação dos conflitos socioambientais no Núcleo Massambaba do Parque Estadual da Costa do Sol. Desenvolvimento e Meio Ambiente, 44. https://doi.org/10.5380/dma.v44i0.54922

Silva Gomes ML, Silva Rebouças L, Varela de Melo GC, Monteiro da Silva MN, dos Santos SM (2019) Evidenciação dos custos e investimentos ambientais das empresas do setor de químicos da B3. Amazônia, Organizações e Sustentabilidade, 8(1). Gale OneFile: Informe Académico, link.gale.com/apps/doc/A598194939/IFME?u=anon~9c714847&sid=googleScholar&xid=1adc56f1. Accessed 3 Aug. 2022

Sulaiman SN, Aledo A (2016) Desastres naturais: convivência com o risco. Estudos Avançados 30:11–23. https://doi.org/10.1590/S0103-40142016.30880003

Thompson F, de Oliveira BC, Cordeiro MC, Masi, BP, Rangel TP, Paz P, ... de Rezende CE (2020) Severe impacts of the Brumadinho dam failure (Minas Gerais, Brazil) on the water quality of the Paraopeba River. Sci Total Environ 705:135914. https://doi.org/10.1016/j.scitotenv.2019.135914

Turner MD (2004) Political ecology and the moral dimensions of “resource conflicts”: the case of farmer–herder conflicts in the Sahel. Polit Geogr 23(7):863–889. https://doi.org/10.1016/j.polgeo.2004.05.009

Urquiza Aguilera AH, de Oliveira Rocha A (2019) The Environmental Disaster of Mariana and the Krenaks of the Doce River. Veredas do Direito, 16:191. https://doi.org/10.18623/rvd.v16i35.1507

Viégas RN (2009) Environmental conflicts and material and symbolical struggles. Desenvolvimento E Meio Ambiente, 19. https://doi.org/10.5380/dma.v19i0.13564

Sergio Ricardo Da Silveira Barros is Associate Professor at the Department of Geoenvironmental Analysis at the Institute of Geosciences at UFF Graduate in Economics and Doctorate in Geography (UFF) emphasis in Territorial Planning and Environmental Planning; master’s in environmental science (UFF); Vice Coordinator of the Undergraduate Course in Environmental Science and Coordinator of REMADS-UFF Network for the Environment and Sustainable Development; Professor at the Doctoral Program in Sustainable Management Systems (PPSIG UFF); Professor at the Master Program on Management Systems (MSG UFF); Participates in the UFF/CNPq Research Groups in Coastal Management at the Geosciences Institute

Sergio Luiz Braga França is Doctor in Civil Engineering (UFF), emphasis in Management, Production, Quality, and Sustainable Development; master’s in civil engineering (UFF); Graduate in Civil Engineer (UFF); Professor at the Doctoral Program in Sustainable Management Systems (PPSIG UFF); Professor at the Master Program on Management Systems (MSG UFF); Occupational Safety Engineer (UFF); Coordinator in the Innovation Center and Technology for Sustainability (NITS/UFF); Coordinator of MBA LATEC/UFF; Reviewer of journals and academic events; Researcher and Consultant at NITS/LATEC/UFF.

Chapter 5
The Governance of "Social Arrangements for Material Reinsertion on New Cycles of Use": A Survey on the Case of the Cooperativa Popular Amigos Do Meio Ambiente (COOPAMA)—Environment Friends State Cooperative

Ligia Moreira, Lucia Cristina Dantas Shpielman, Micheli Santos De Souza, and Raquel De Oliveira Moreira

Abstract This chapter presents a case study of Cooperativa Popular Amigos Do Meio Ambiente (COOPAMA), a collector's cooperative located in Brazil, illustrating the Environmental, Social, and Governance (EGS) principles in its organization. Founded in 2004 in Complexo do Jacarezinho, Rio de Janeiro, as a response to the lack of opportunities in the conventional market, COOPAMA stands out in the management of recyclable waste on a large scale. With 86 members, it plays an important role in collecting and recycling waste, establishing partnerships with municipal collections and companies, and installing ecobarriers to prevent river disposal. In addition, it promotes cultural, social and training programs, positively impacting the environment and local social and economic development. This approach aligns with ESG principles, exemplifying the effective integration of these principles into operations and their broader positive impact.

Keywords Recyclable waste · Cooperative · Collectors · Recycling · ESG

L. Moreira · L. C. D. Shpielman (✉) · M. S. De Souza · R. De Oliveira Moreira
Management Systems, Fluminense Federal University, Niterói, Brazil
e-mail: lcgd.shp@gmail.com

R. De Oliveira Moreira
e-mail: raquel_moreira@id.uff.br

M. J. Meiriño and O. L. G. Quelhas (eds.), *Critical Factors Associated with Environmental, Social and Governance Management in Organizations*, Accounting, Finance, Sustainability, Governance & Fraud: Theory and Application, https://doi.org/10.1007/978-981-96-1725-8_5

5.1 Introduction

The contemporary evolution of society has been accompanied by a consumerist culture, in which it has sought to prioritize economic growth, which, consequently, has exponentially increased waste generation. However, the lack of awareness about the destination and/or final disposal of waste raises concerns about the impacts generated when they are not disposed of properly. It is worth mentioning that most urban solid waste (MSW) is of industrial, domestic, hospital origin, among others. The pressure suffered on the environment is directly related to economic, social, cultural, political and anthropogenic factors, as well as significantly affecting human health.

The search for improving Environmental, Social, and Governance pillars has intensified within organizations, which have demonstrated a closer look at the environmental aspect. Considering the above, it is noteworthy that the emergence of the term Environmental, Social, and Governance (ESG) occurred from a joint initiative of financial organizations with the purpose of developing guidelines and recommendations on how to better integrate ESG (ECCLES et al., 2019).

In this context, the relevance of the fundamental environmental services that Catadores Cooperatives provide is highlighted, as well as being survival alternatives for families in vulnerable situations. Cooperatives have generated opportunities for cities to develop cleaner, in addition to contributing to reducing the amount of waste sent to landfills. Considering the relevance of cooperatives in environmental, social and economic aspects, it is noteworthy that the application of ESG principles allows them to increase transparency, operational efficiency and reduce risks. It can also achieve more sustainable performance, implement social responsibility policies, improving the quality of life of the surrounding communities (Saueressig et al., 2021).

5.2 The Chapter's Role in the Book Structure

There are many definitions for the term *Environmental, Social, and Governance* (ESG), but broadly it can be understood as a compromise that an organization must have with the society, environment, and the governance of its operation. In Latin America, especially, this theme is still in the maturing phase and only a few organizations are actually able to develop ESG practices integrated into their management system. (Anholon, 2016).

Following this book's study line, this chapter aims to describe and analyze the governance indicators of a recyclable materials collectors' association, integrating the recyclable waste management theme to the ESG, aligned with continuous improvement.

This study has highlighted some ESG indicators: Organizational responsibility, ethics, conduct and conflict of interest, transparency, capacity building, management, inspection, and audit.

Therefore, the possibility of applying those practices even to the simplest organizational structures is evident since the essence of the ESG concepts is about providing a more sustainable world to guarantee the future of future generations.

5.3 History of the Theme Development

The ESG theme emerged in 2005 in the *"Who Cares Win"* report. This publication results from a meeting with the then-UN General Officer and CEOS from large financial institutions on integrating social, environmental, and governance factors in the capital market.

Currently, this theme has stood out in the organizational and academic environment. It is noticed by the growing number of organizations that assume their compromise for a better world and by the recent research reports published in the last few years. In addition, the values related to sustainable development are being institutionalized to a greater or lesser extent in various countries by the media, social movements, environmentalists, and governments (Barbieri et al. 2010; Debeljak et al 2011; Frynas e Stephens, 2015). Despite this importance, however, Kirat (2015) argues that there is still no broad and precise consensus about the real meaning of the term, giving space for new research and debates in the academic community.

The National Solid Waste Policy (PNRS), published through the Brazilian Law nº 12.305, in 2010, established several instruments for its implementation, the selective collection is among them. The Municipal Plans for Integrated Waste Management (PMGIRS) elaborated on predicting this activity. It is worth mentioning that waste is a strong methane gas emitter (CH_4), and it presents a 21 times global warming potential, bigger than carbon dioxide gas (CO_2), the reason why PMGIRS should consider the need to reduce these greenhouse gases, data extracted from PMGIRS, in 2021 of the Rio de Janeiro State.

In this context, selective collection helps to reduce the emission of those gases, restricting the volume of waste sent to landfills or dumped incorrectly into the environment. Besides, the implementation of selective collection implies the partnership with cooperatives or associations of recyclable material pickers, and for the PNRS, these associations have great relevance in the selective collection process.

Lima (2018) signaled that amidst many difficulties in entering and remaining in the job market, the solidarity economy movements, the cooperatives, and the organization of workers in associations continue to appear as an alternative in the direction of a more egalitarian society based on the inclusion of workers who, either because of the demands of the market or a historical process of marginalization, are unemployed.

Nowadays, issues related to the environment and its preservation have acquired greater visibility since the degradation is increasingly evident, compromising economic and social development. Therefore, it is necessary to design solutions that have the essence of committing to the environmental, social, and financial aspects, aiming at equity among the parties.

Notably, human activities have been generating enormous impacts on the environment, in the exacerbated exploitation of natural resources or the irregular disposal of waste generated. In this sense, organizations have sought ecologically correct solutions aiming to guarantee their permanence in a regenerative way, contributing to the environment and society, associated with the economic growth that has been driving the search for conscious and sustainable business.

Given the concern with sustainable development, the term ESG is being used more frequently. In addition, it serves as a parameter to demonstrate how committed an organization is to environmental, social, and governance issues. Thus, the ESG approach brings in points necessary to sustain a conscious and prosperous business.

Corporate governance can be considered a tool that aims to provide better transparency, discipline, and responsibility of company managers toward shareholders, seeking, therefore, to minimize the discordance of clarifications between the company and the various actors involved since the relevance and transparency of the information, related to a management policy are essential in its development. In Brazil, Law nº10.303, enacted in October 2001, decreased shareholder concentration in the national capital market, offering greater accessibility to small investors. The changes implemented in this Law are associated with measures regulated by Bovespa, Comissão de Valores Imobiliários (CVM) [Securities and Exchange Commission (CVM)] and the Central Bank (Bacen), symbolize a very relevant milestone in the progress of governance in the country. Brazilian and international experience have shown that investors prefer companies with higher levels of governance since they reflect a minimum risk to capital (Vieira; Mendes, 2006).

5.4 Conceptual Structure

Cooperative governance is the set of processes, policies, laws, and regulations that underlie the way a cooperative is directed, managed, and controlled, enabling its members to ensure the execution of organizational objectives. Therefore, by acting according to the principles of cooperative governance, the cooperative which follows these principles will be more likely to achieve the loyalty of its members, as well as to institutionalize an environment of trust for all those involved in the cooperative's activities over time (Guilherme; Pinto; Kochhann Reisdorfer, 2015).

According to Valverde's study (2021), recycling cooperatives play an indispensable role in reducing the impact of solid urban waste on the environment, besides providing income generation and social inclusion. Therefore, it is understood that good cooperative governance ensures fairness of treatment, legal compliance, transparency, and responsible accountability to the members so that the cooperative society has its value and perpetuity increased.

In recent years, the generation of municipal solid waste (MSW) has increased significantly, and recycling is an alternative to reduce the amount to be managed. Thus, waste picker cooperatives and the recycling industry can contribute to mitigating the effect of increased MSW generation. The cooperatives connect the final consumer and the recycling industry, contributing to the collection, separation, and processing of some materials, but do not have direct contact with the recycling industry. Without a doubt, the importance of the activities of the collector's cooperatives lies in the impact they have on the conservation of natural resources by stimulating their multiple use, in the generation of employment and income for vulnerable populations (Saueressig; Sellitto; Kadel, 2021).

In Brazil, the National Policy for Solid Waste (PNRS) recognizes the activity of the collectors of recyclable materials, having included in its objectives and encouraging the creation and development of cooperatives or other types of association for the purpose of promoting social inclusion and economic improvement for these workers, recognizing them as key players in the GRSU. For this reason, government agencies should evaluate the environmental impacts of different types of waste disposal throughout the entire life cycle and establish sustainable household waste recycling systems under the guidance of academic research. The influential factors behind residents' attitudes and willingness to recycle must be seriously considered in the policy-making process (Shi; Zhou; Zhang, 2021).

In conjunction with this, to correspond to the category of social inclusion, the formalization of the collectors' activity must be thought of more broadly, articulated with a policy that ensures stable channels of supply and access to recyclable materials for the cooperatives, since the simple concession of physical structure and machinery, a fundamental step, does not necessarily correspond to the economic viability of the enterprise and to obtaining a minimum income for its members (LIMA, 2018).

From the information presented, it is understandable that the Waste Pickers Organizations are the fundamental link in the integrated management of municipal solid waste and play a key social, economic, and environmental role. Therefore, the main activity performed by the Pickers' Organizations is to insert materials in the productive cycle, promoting a circular economy but, due to financial dependence on the public sector as the primary economic source and the difficulties of self-management, these organizations do not reach a competitive position in the recycling market (ROMEIRO, 2021); (Tong; Huynh; Khong, 2021).

With the growth of the cooperative movement in Brazil and in the world, cooperative governance is increasingly becoming a strategic management modality within cooperative organizations, whose essence is self-management and a peculiar way of distributing ownership rights, decision-making power, and residual gains. Furthermore, the analyses indicate that the practices of cooperative governance in these organizations are still very scarce, especially regarding internal oversight and management of people and conduct (VALVERDE, 2021).

According to Mendonça and Filho (2004), in Third Sector organizations sustained with resources from dispersed donors, there is a strong assumption that managers will allocate resources as efficiently as possible, but not assured by extensive monitoring by principals. In this case, governance mechanisms to reduce conflict of interests between donors and managers may be a typical donor collective action problem.

Based on the research of Rezende, Facure, and Dalmácio (2009), carried out with basis on the assumptions of corporate governance, IBGC recommendations, legal rules, and research suggestions, the Football Clubs Corporate Governance Index

(IGCCF) was elaborated and it was verified that the general level of adoption of these practices, in the Brazilian soccer clubs, is extremely low (average use of only 28.7% of the total score that could be reached).

Also, in Valverde's study (2021), the cooperative governance practices are shown as a path to ensure the achievement of corporate and social objectives through a strategic management model, so the intention is to verify which practices are applied to the management reality of the cooperatives analyzed. The study also reported that the feeling of belonging within the organization is indispensable for the cooperative members' commitment, and it is noticeable in smaller organizations such as the ones analyzed. Another point that contributes to the absence of cooperative governance practices in these institutions is the lack of training and specialization of managers to institute them since they would lead the organizations to an increasing structural and operational complexity.

Many types of research have emphasized the strategic management of Corporate Social Responsibility (CSR). However, there are still gaps in how CSR can be effectively integrated into existing business processes. Previous literature on CSR has established the need for CSR practices, however, the current challenge is to integrate CSR as part of business processes. In this context, an integrated management systems approach to corporate social responsibility, the article by Asif et al. (2013), focused on how organizations can integrate CSR into business processes by identifying the fundamental principles that are inherent in CSR practices and management systems adapted to meet the needs of a variety of stakeholders.

5.5 Coopama Cooperative Case Study

Box 1 Case Study

COOPAMA. Available at: < https://www.coopamacoop.com/sobre > **. Accessed on: 27 feb. 2022.**

The Environment Friends State Cooperative (COOPAMA) case study, located in the municipality of Rio de Janeiro, in the Maria da Graça neighborhood, will be presented. It started its activities in January 2004 by a group of workers, residents of the Complexo do Jacarezinho, who, with no perspective of returning to the market, opted to organize themselves through collective work. The Jacarezinho Community is in the municipality of Rio de Janeiro, between the neighborhoods of Engenho Novo, Jacaré, and Del Castilho. The Jacarezinho Community is in the city of Rio de Janeiro, between the neighborhoods of Engenho Novo, Jacaré, and Del Castilho.

COOPAMA works with various types of recyclable residues, carrying out the pickup, sorting, recovery, and destination of solid recyclable waste on a large scale and in different contexts, including several partnerships: With the official selective collection from the Rio de Janeiro city and with big companies that generate waste. Acting in urban areas during events and cleaning rivers by installing ecobarriers to retain the disposed waste and providing consultancy and lectures on waste management in various organizations.

The cooperative currently has 86 members, at the organizational level are the CEO, Production Director, Director of External Services, Environmental Technical, and Administrative. It has spaces for the reception, segregation, storage, and compaction of residues, in three covered warehouses with an area of 1000 m2 of the former Hitachi Line factory. Its goal is to receive the solid residues that can be reused or recycled, reducing the negative environmental impacts linked to the misused destination and/or disposal of it, contributing to the extension of products and packages life's cycle through the collection, separation, and supply of secondary raw material for the industry, and also collaborating with the local economy by generating income for the cooperative members.

According to COOPAMA's website, the cooperative is composed of people united with democratic management with the mission of satisfying economic, social, and cultural needs and aspirations through sustainability. Over the years, COOPAMA has matured its administrative and operational procedures and has incessantly sought to train its work team. It has received recognition in the market and is currently a partner of companies, research institutions, and environmental agencies. Throughout these years, it has evolved with recyclable residues generated by the principal productive chains, placing itself in the market as a fundamental player to meet the needs of companies and society.

The comparative analysis of the information was based on the literature review from 2005 to 2021 and a face-to-face interview held on 09/17/2021 at COOPAMA's headquarters, a virtual meeting held on 02/21/2022, through the Google Meet website, lasting two hours, a telephone interview, and also the COOPAMA website was consulted.

The information highlighted in this study tends to corroborate the relevance of the service provided by the cooperatives, especially COOPAMA, which shortly may expand its network of partners, offering mutual benefits.

In the interviews conducted with COOPAMA representatives, the questions were based on corporate governance indicators.

As for corporate responsibility, we highlight the monthly meetings held with the fiscal council, in which time is arranged to record discussions and future reference. Periodic meetings are also held with the board, management, and members.

According to one of the directors, the cooperative has similar activities to a company.

The participation of the members in the assembly meetings provides unity and partnership to generate new opportunities and social development, showing the commitment to the relationship among the class.

Ethics, conduct, and conflict of interests are part of the cooperative's internal statute, which reveals a more mature organizational model, proposed to solve eventual problems/conflicts through interpersonal behavior, with a protective internal bylaw.

According to the director, issues are solved by the person responsible for the area, who goes to the location, and if not managed, then members are addressed to the cooperative direction to talk.

As for transparency of information, the cooperative has good adherence to creating groups through the social network "WhatsApp," usually organized by project, for which all members must have a cell phone. According to the interviewees, periodic meetings happen monthly or bimonthly, with the members to follow up on the development of the activities.

About the capacitation, the interviewee informed that there are constant upskilling courses for the workers. The topics vary from the cooperative movement; use of PPE; basic English; training for the cooperative members who use compacting, fragmenting, or others equipment; carpentry workshops for restoration, furniture creation, reusing donated furniture or wood, etc.

Activities deliberations are made at meetings by the board and transmitted to those responsible for each sector or project. The fundamental requirements for choosing candidates for board positions depend on the interest in participating in the cooperative's internal activities, being active, and wanting to be trained. Thus, no minimum requirements are set for the performance of these actors in their respective statutory functions.

The board of directors performs the control and monitoring of the resources through an Excel spreadsheet program. The cooperative is developing its indicators, preparing monthly reports with information about incoming and outgoing waste.

The interviewee informed that inspections are not frequent. During his tenure, the cooperative only had one inspection visit. He also informed that it has an operating license from the Municipal Department of Environment (SMAC), registration in the Waste Manifest System of the State Environmental Institute, and the Federal Technical Register of Potentially Polluting Activities and/or Users of Environmental Resources (CTF/APP) of the Brazilian Institute of Environment (IBAMA). Besides, it has a Medical Control Program of Occupational Health—PCMSO and a Program for the Prevention of Environmental Risks.

COOPAMA's organizational identity is composed of mission, vision, and value. It is guided by the principles of a cooperative and solidarity economy, having the purpose of being "a collective of recyclable material workers" who benefit from work and income generation, promoting the reuse of products, saving natural resources, and benefiting the society. COOPAMA's values are Sustainability, Social Inclusion, Cooperativism, Empathy, and Organization. (Jokubowicz, Junior, 2018).

COOPAMA has as partners' large companies and organizations that adopt sustainability, are responsible for social welfare, and understand that the correct management of waste benefits society and the environment. The organization of cooperatives brings these workers better working conditions and income. In Brazil, the majority of waste pickers still work in the streets and garbage dumps of the country, autonomously and in precarious working conditions. A cooperative's goal is to generate job opportunities, income, and social inclusion for the collectors, reducing the vulnerability situation of needy communities, providing a permanent option of work and for this population.

COOPAMA works with multiple types of recyclable residues, such as plastic, cardboard, paper, glass, ferrous and non-ferrous metals, electro-electronics, used cooking oil, Styrofoam, and furniture. The cooperative has some partnerships such as the selective collection of the Municipal Company of Urban Cleaning (COMLURB) and other large companies: Via Varejo, Petrobrás, Furnas, INCA, Labor Ministry, Public Ministry, Norte Shopping, Renner, Plaza Shopping, SESC, SENAC, Hospital Marcílio Dias, Boticário, among others. Currently, approximately 120 tons of residues are no longer discarded monthly in the environment and are sold to industries, separated, and weighed at the cooperative, generating around R$160,000.00/month. The average daily remuneration of the collectors is R$100 reais, approximately two minimum wages per month, working only on weekdays. The director also explained the seasonality of sales, which increases between the months of January and April and on festive dates.

Its exemplary performance with non-hazardous waste, including electronic waste, vegetable oil, Styrofoam, and wood, and the development of several social projects such as woodworking courses to train the cooperative members and other projects. That made COOPAMA stand out in the field of its activity attracting the attention of academic students with the goal of evaluating the integration of ESG indicators within the cooperative's organizational structure, as we see below.

I. Commitment to environmental issues

The cooperative participates in external projects such as the "Ecobarriers Project," in which the members build barriers with plastic drums and nets in rivers for the retention of incorrectly discarded residues. At the location itself, the residues that do not present a high degradation index are removed and taken to the cooperative, and the rest is sent to the sanitary landfill.

II. Commitment to social issues

Cultural and Social Programs and Partnerships. Its pillar programs, highlighted in the social scope, aim to increase the community's capacity to continuously improve the quality of life:

Cult and Art Project

The project uses recyclable waste as a tool to raise awareness among different segments of society (mainly children and young people) about the importance of this material as a tool and infrastructure for many professionals, such as waste pickers, artisans, historians, and environmentalists.

An art gallery was built on the site, in partnership with the Cult and Art Institute, to exhibit the art created from recyclable waste. This space is open to the whole community allowing a wide interaction between the cooperative and the neighborhood, besides offering the opportunity to show how its products and services contribute to the well-being of individuals and the general human development.

We highlight the promotion of educational visits, which help to spread the word about the work done in the cooperative. Among the many pieces in the permanent exhibition, "The garbage that gets transformed" is a representation of the reality of mangroves, beaches, and rivers without proper waste management. This space is managed by the leaders of the cooperative, the location is open to visitors, and receives students from high schools and universities for environmental education activities.

Capacity Building in Carpentry

Besides capacitating the cooperative members, the carpentry project intends to reuse the donated wooden furniture as raw material for the students' training.

Projects with Vulnerable Groups

It has a public policy specialist to support the management of social projects, Ms. Nilza Gomes. Currently, some projects are being developed for these groups, such as workshops, filming, and the production of educational videos.

Women in Cooperative Movement Workshops

These workshops are designed to empower female waste pickers. They were elaborated by the waste pickers' daughters, who today live in other environments, such as universities, and bring to these meetings their knowledge. In this project, there will be workshops, conversations, and debates. Also, content relevant to their daily experiences and practices will be sent via WhatsApp, creating tools that can help these women, who live on the margins of society, to develop new paths in their routines.

Women in Urban Quilombos

It is a film project that portrays the favela as a big house and the houses that the women reside in are considered rooms inside of this bigger house. According to the interviewee, Ms. Nilza Gomes, the slum interacts in a different way and sequence from the economically structured society, with a guiding principle of its routines based on solidarity among its residents. The video aims to bring the reality of the slums so that interventions in those communities are based on the daily life and experiences of its residents.

III. Governance commitment

From the technical visit made to the cooperative, we verified the commitment that the then CEO, Mr. Luiz has collaborating with sustainable development, demonstrating a concern with the continuous improvement of COOPAMA. Therefore, he presented problems within the cooperative, such as the lack of training of the members, and according to him, they must be qualified for being prepared for a possible absence of a member of the cooperative's organizational level. The CEO's concern for the education and health of his members is well known.

It is worth mentioning that COOPAMA has partnerships with large, medium, and small organizations. Besides being the only cooperative in the state of Rio de Janeiro that receives Styrofoam, a residue that presents a huge environmental impact because its decomposition is considered indefinite.

5.6 Indication of How the Conceptual Framework Will Be Incorporated into the ESG Development, Deployment, and Monitoring Process

Based on the literature studied, following Asif et al's (2013) model, it presents an integration of CSR practices into existing business organization processes. The model highlights the concept of simultaneous top-down and bottom-up approaches simultaneously. According to Anholon et al (2016), the top-down approach focuses on identifying stakeholder needs and also integrating CSR practices with internal management systems.

The bottom-up approach, on the other hand, provides a system whereby organizations can interact with the community to understand better how the organization's activities influence them, how living standards can improve, which are the priority projects to support, and which indicators need to be managed in order to identify the achievement of objectives.

According to Asif et al (2013), the model was based on the Plan-Do-Check-Act (PDCA) cycle of continuous improvement, which provides organizations with a meta-routine for ongoing evolution in the social, environmental, economic, and other stakeholder dimensions. The meta-routine goes beyond daily operations and emphasizes the importance of learning, innovating, and developing institutional knowledge.

In the model presented by Asif (2013), activities are related to a top-down and bottom-up structure. In top-down activities, the organization must carry out a survey in order to identify and prioritize the needs of its stakeholders. In the top-down approach, the identified requirements also need to be assessed against parameters such as quality, occupational health and safety, volunteer initiatives, customer satisfaction, environmental impacts, equity issues, among others. Therefore, provides a systematic approach for translating ESG/governance strategic goals and integrating the different stakeholder requirements into business processes.

In bottom-up activities, the organization must relate to the communities and identify their needs. Indicators linked to the community needs are defined, seeking to integrate these results with the management systems in place in the organization (Asif et al, 2013).

In the Do phase, the organization's management strives to make ESG practices part of the daily organizing routine, preparing manuals, procedures, work instructions, training, capacity building, among others (Asif et al, 2013).

The Check phase is the measurement and evaluation stage, which can be carried out by evaluating indicators, audits, and benchmarking. The results arising from this review will be spread to all stakeholders involved, either by publishing the reports or another communication tool. In the Act phase, communication is a pillar. The results of the indicators, innovative processes, the knowledge obtained, and new needs observed feedback in the process, are defined as a cycle of continuous improvement (Asif et al, 2013).

Facing an organization development process, it is worth pointing out that the inter-relations of the agents involved are of utmost importance for the productive and operational constitution. In this sense, cooperatives also require a considerable level of governance to reconcile and align better the interests of those involved in their structure. Governance practices are an essential guiding instrument to adapt and guide the efficient functionality of cooperative organizations, which enable their sustainability and continuous improvement (Costa; Watanabe, 2018).

The governance actions were noticeable in several aspects of the cooperative. As far as corporate responsibility is concerned, one can see in the day-to-day management of the organization the actions already described in its statutes and bylaws.

In this context, as a suggestion for improving governance in cooperatives, here are some ESG indicators observed in this chapter.

In the cooperative's activities planning, periodic meetings with their boards, board members, and cooperative members should be forecast in an annual schedule of meetings. It is crucial to hold monthly meetings with the fiscal council, as it is part of the governance structure and acts in the surveillance of the administrators' acts. It is also important to take minutes of the meetings to record the discussions and future references.

The meetings have been assertively present in such a way that the members tend to understand better the progress of the cooperative, the management structure and tend to get involved more effectively in the productive system of corporate management, resulting in better governance practices in these organizations and consolidation of the participation principle by the members. Also offering greater access to information, this channel provides resources to the cooperative members, building opinions and argumentation about the topics on the agenda (Valverde, 2021).

The structuring of management and governance processes is indispensable to harmonizing the conflicts of interest that might appear. In cooperatives, the collective must be above the members' individual will. Adopting measures that effectively ensure and promote balance among the organizations' stakeholders is essential. Optimizing the interactions among members, democratizing relationships, and improving the ways relations take place in the internal environment of organizations are shown to be necessary and effective governance measures to reduce or curb internal conflicts (Ferreira, 2022; Valverde, 2021).

The internal statute of the cooperatives reveals a more developed organizational system, proposed to solve eventual conflicts based on interpersonal conduct, with an internal statute that protects them, where eventual problems and issues are solved. If necessary, the person responsible for coordinating the activities of each area or the management solves problems that may appear.

Transparency is necessary to ensure that stakeholders have confidence in the decision-making process (Vieira & Barreto, 2019). The information needs to be balanced and of quality, addressing both positive and negative issues in order to provide stakeholders a correct perception of the cooperative's reality (Sistema Ocb, 2010). Internal communications can be carried out in addition to social networks or by phone calls. These practices favor stakeholder access to information and provide transparency within the cooperatives' management.

The training of cooperative members is fundamental in the governance process of the cooperatives. Qualifying the performance of the collectors and strengthening their participation in the recycling chain is to collaborate for the sustainable development of the country. According to Valverde (2021), the cooperative board of directors plays a fundamental role and is considered the management and administration body, responsible for producing results, maintaining the social and financial balance of the business, using tools for planning, organization, and control. In this way, for the best operation, the direction of the organ needs technically trained managers.

One of the pillars for the management of cooperatives is the monitoring and control system, and it would be no different for recycling cooperatives. However, external inspection practices are not yet recurrent in the reality of these organizations. As for the internal control of the licensing, occupational health, and environmental risk prevention programs, we suggest keeping this data updated, organized in spreadsheets, controlled by a professional responsible for each activity.

5.7 Summary of Main Ideas and Concepts

- ESG, the acronym created in 2004 stands for environmental, social, and governance.
- Environmental (E), are practices related to the conservation of the environment.

- Social (S), concerns a company's relationship with the society around it.
- Governance (G), best practices for managing a company.
- About the collectors' cooperative governance, it was observed that there is a clear search for change in the current model, adapting to a new context, that is, reviewing its structure and processes, standing out in the way they operate, valuing the work of each one within COOPAMA, and organizing themselves so that the proper functioning of the cooperative generates satisfactory results for the members, as well as for society and the environment.
- Who Cares Wins. Report where the Global Compact was published in 2004, the result of a meeting held by an initiative led by the UN with 20 financial institutions from 9 different countries—including Brazil.
- A recycling cooperative is an institution that performs a set of actions, directly or indirectly, in the stages of collection, transport, transshipment, treatment, and final environmental destination of discarded residues.
- The waste picker, is an urban worker who collects recyclable solid waste such as cardboard, aluminum, glass, and others.
- Cooperativism is an economic and social movement, between people, in which cooperation is based on the participation of members, in economic activities with a view to achieving the common good.
- CEOs, stands for Chief Executive Officer. It is the person with the highest authority in the operational hierarchy of an organization. He or she is responsible for the company's strategies and vision.

5.8 Questions for Reflection and Study

(1) What are the main ESG philosophy projects developed by the organization studied?
(2) Identify in the Case study evidence of recognition of governance in the organization?
(3) Based on the integration model proposed by Asif et al (2013), can it be said that COOPAMA has a management system that integrates the pillars of the ESG concept into its daily activities?

5.9 Cases to Be Solved

(1) A collectors' cooperative was visited by inspectors for sanitary inspection. Most of the questions raised by the inspectors were answered, and the nonconformities were later sent in a report with a deadline to attend to them. Can this surveillance become a positive action for the cooperative? Why?
(2) A waste picker cooperative needs ESG indicators to analyze its performance. Observing the studied cooperative, no indicators for governance management were found. Which indicators could be suggested to supply this demand?

References

Anholon R, Quelhas OLG, Leal Filho W, Pinto J De S, Feher A (2016) Assessing corporate social responsibility concepts used by a Brazilian manufacturer of airplanes: a case study at Embraer. J Clean Prod, [s. l.], ano 2016 135:740–749

Asif M et al (2013) An integrated management systems approach to corporate social responsibility. J Clean Prod 56:7–17

Barbieri JC et al (2010) Innovation, and sustainability: new models and propositions. Bus Adm J 50(2):146–154

Costa, E. S.; Watanabe, M. Governança E Cooperativa: Nível De Governança Nas Cooperativas De Santa Catarina. II Jornada Nacional De Desenvolvimento E Políticas Públicas. Labor And Development In Latin America. Universidade Do Extremo Sul Catarinense (UNESC) September 2018

Debeljak J, Krkač K, Bušljeta Banks I (2011) Acquiring CSR practices: from deception to authenticity. Soc Responsib J 7(1):5–22. https://doi.org/10.1108/17471111111114503

Eccles RG, Lee L-E, Stroehle JC (2019) The social origins of ESG: an analysis of innovest and KLD. Organization & Environment, 33(4):575–596. https://doi.org/10.1177/1086026619888994 (Original work published 2020)

Ferreira MP, Gomes JS, Freire FB (2022) A scientometric review of governance and conflicts in family businesses/a scientometric review of governance and conflicts in family businesses. Revista FSA (Centro Universitário Santo Agostinho) 18(10):23–48

Frynas JG, Stephens S (2015) Political corporate social responsibility: reviewing theories and setting new agendas. Int J Manag Rev 17(4):483–509

Guilherme N, Pinto M, Kochhann Reisdorfer VG (2015) Cooperativa 2015 Santa Maria-RS. [S.L: S.N.]

Jakubowicz Junior D (2018) Proposal of a methodology for mapping reverse logistics in waste pickers cooperative

Kirat M (2015) Corporate social responsibility in the oil and gas industry in qatar perceptions and practices. J Public Relat 41(4):438–446

Lima MRP (2018) Paradoxes of formalization: The social inclusion of waste pickers from the case of the closure of the landfill of Jardim Gramacho (RJ). HorizTes Antropológicos 24(50):145–180

Mendonça LR De, Filho CAPM (2004) Governança Nas Organizações Do Terceiro Setor : Considerações Teóricas. R. Adm V 39:302–308

Rezende AJ, Facure CEF, Dalmácio FZ (2009) Práticas De Governança Corporativa Em Organizações Sem Fins Lucrativos. Anais Do Congresso Usp De Controladoria E Contabilidade

Romeiro AR (2021) Civilization, and innovation–why the industrial revolution was a phenomenon dependent on a civilizational trajectory dependent on a civilizational trajectory. Inst Econ Da Unicamp

Saueressig GG, Sellitto MA, Kadel NPD (2021) Cooperativas De Reciclagem No Retorno Dos Resíduos Sólidos Urbanos À Indústria. Revista Em Agronegócio E Meio Ambiente 14(2)

Shi K, Zhou Y, Zhang Z (2021) Mapping the research trends of household waste recycling: a bibliometric analysis. Sustainability (Switzerland) 13(11)

Sistema OCB (2010) Manual De Boas Práticas De Governança Cooperativa. p 44

Tong YD, Huynh TDX, Khong TD (2021) Understanding the role of the informal sector for sustainable development of municipal solid waste management system: a case study in Vietnam. Waste Manag 124:118–127

Valverde GM (2021) Análise Da Governança Cooperativa Nas Cooperativas De Reciclagem De Florianópolis/Sc. Universidade Federal De Santa Catarina Centro Socioeconômico Departamento De Ciências Contábeis Graduation In Accounting Sciences

Vieira JB, Barreto RTDS (2019) Governance, risk management, and integrity. Brasília: Enap. Available: http://repositorio.Enap.Gov.Br/Handle/1/4281

Vieira SP, Mendes AGST (2006) Governança Corporativa: Uma Análise De Sua Evolução E Impactos No Mercado De Capitais Brasileiro. Revista Organizações Em Contexto 2(3):48–67

Ligia Moreira Specialist in Public Finance from Cândido Mendes University. Holds a degree in Accounting Sciences from Estácio de Sá University. and has an extension course in Quality Management from the Federal University of Rio de Janeiro and received the title of Internal Quality Auditor from COPPE Quality. She works in the administrative area of the Brazilian Navy. In the Quality Management area, she has already participated in organizational management committees, taking part in the elaboration of several documents, including strategic plans of the organizations where she served, besides currently being part of the Executive Committee of Risk Management of the Unit where she operates.

Lucia Cristina Dantas Shpielman Specialist in Hospital Administration from Centro Universitário São Camilo. She has a degree in Business Administration from Faculdades Integradas Moacyr Sreder Bastos. Served as an Administration Supervisor at the National Cancer Institute José Alencar Gomes da Silva (INCA), in several areas, during the period from 2004 to 2019. Main activities: waste management, building supervision, contract supervision, online teaching.

Micheli Santos De Souza Specialist in Labor Safety Engineering from Universidade Federal Fluminense. Has a degree in Production Engineering from Universidade do Estado do Rio de Janeiro. She is currently a Science & Technology Analyst at the National Cancer Institute José Alencar Gomes da Silva (INCA), working mainly on the following topics: quality, safety, and environment.

Raquel De Oliveira Moreira Specialist in Planning and Management of Hydric Resources by Cândido Mendes University, specialist in Management of Solid Waste and Effluents by Faculdade Única. She has a degree in Environmental and Sanitary Engineering from Universidade Estácio de Sá. Also worked as Coordinator of Environmental Projects in the Municipal Secretary of Environment and Urbanism of Itaboraí, developing projects, organizing events, promoting environmental education activities with emphasis on solid waste management both in school environments and in public agencies.

Part IV
Governance Perspective

Chapter 6
An Analysis of Corporate Social Responsibility Initiatives: Studies of Public and Private Companies

Mirian Picinini Méxas and André Cantareli da Silva

Abstract Sustainable development is imperative for the advancement of the planet itself. Corporate actions initiated in the 1950s in some countries and now undertaken by many worldwide demonstrate corporations' concern for environmental issues through Corporate Social Responsibility, undertaking various initiatives. This chapter analyzes the main sustainability actions of public and private organizations in Brazil and South America. The methodology employed involved documentary research and case studies of four organizations established in the country, two of private initiative and two of public initiative. The research encompassed articles, publications, digital documents, and a study on each organization. The selected companies are from distinct sectors: A cosmetics company, Natura S.A.; a construction company, Saint-Gobain Brazil; a sanitation company, SABESP—São Paulo Basic Sanitation Company; a nuclear materials company, INB—Nuclear Industries of Brazil. Corporate Social Responsibility initiatives were analyzed and presented descriptively in a box format. It can be observed that the organizations have various sustainable development actions, such as economic growth associated with social and environmental promotion, respect and integrity in relations with employees, supplier management with ecological criteria, and environmental monitoring through dialogue with society, thereby contributing to the planet's environment.

Keywords Corporate · Initiatives · Responsibility · Sustainability · Transparency

M. P. Méxas
Civil Engineering, Actuarial Sciences, Federal Fluminense University, Niterói, Brasil
e-mail: mirian_mexas@id.uff.br

A. C. da Silva (✉)
Civil Engineering, Department of Accounting Sciences, Federal Fluminense University, Niterói, Brasil
e-mail: andrecantareli@id.uff.br

M. J. Meiriño and O. L. G. Quelhas (eds.), *Critical Factors Associated with Environmental, Social and Governance Management in Organizations*, Accounting, Finance, Sustainability, Governance & Fraud: Theory and Application, https://doi.org/10.1007/978-981-96-1725-8_6

6.1 Introduction

Indisputably, the world is experiencing the most crucial moments of a sustainable decision-making process that must actually guarantee the continuity of life on the planet in a bearable way. Only in this way, with true environmental awareness, will life continue in full conditions (ELKINGTON, 2018).

This perspective has guided organizations around the planet when they practice Corporate Social Responsibility. Within this context and the book's main theme, this chapter analyzes Corporate Social Responsibility–CSR initiatives, studying Brazilian and foreign companies, public and private.

It also seeks to study the formation of the concept of CSR, sustainable development and a brief history of the construction of these concepts, demonstrating that since the 1950s (University Of Iowa, 2022), authors, businesspeople, organizations have already sought to guide collective actions in the corporate world, in the middle of the century that became known as the century of the 2nd industrial revolution (Stock et al., 2018).

Globalization is the phenomenon that, stemming from the modernization of communication systems identified in the final two decades of the twentieth century, has rendered the world more interconnected and, consequently, more informed about diverse realities, preserved environmental processes, more polluting systems, and organizations inadequately aligned with environmental and social aspects and social disparities.

Therefore, this chapter analyzes Corporate Social Responsibility initiatives in public and private organizations in Brazil and South America. Organizations from various sectors were selected to demonstrate effective sustainability actions.

6.2 Role of the Chapter in the Book's Structure

Faced with the proposal of a book whose title "Critical Factors for ESG Environmental, Social and Governance Management Implementation in the Organizations" demonstrates the concern of its organizers and authors, with a relevant theme, this chapter aims to present the realization of Social Responsibility Corporate–CSR in public and private organizations, building for this a brief history of the creation of the term and its applicability in Organizations, as a specific objective.

In this way, CSR has become essential to Organizations in a globalized world with economic, social, and environmental guidelines that are necessary for all segments. Some actions in this direction can be urgently considered, as in John Elkington's Triple Bottom Line, which is more current than ever, even though it is a theory developed more than 25 years ago.

The author himself writes that the model suggested by the expression should be rethought, proposing what he called a recall in management concepts, as occurs in industries. The Triple Bottom Line is "a sustainability framework that examines a company's social, environmental and economic impact" (Elkington, 2018). The expression became also known as the sustainability tripod.

With this principle, there is a need for balance in ESG actions because when analyzing the three pillars of sustainability in detail, it is observed that there is no way to prioritize without harming the other. Moreover, how can investments be made without affecting the environment and society? Every enterprise, however small, will always generate impacts, whether major or minor, reversible or irreversible, passive or not, for mitigation.

Thus, this chapter reviews terms, expressions, and meanings more applied to Corporate Social Responsibility and the observation and analysis of CSR cases in the public and private sectors.

6.3 Theme Development History

The thought of Corporate or Business Social Responsibility can be considered as its starting point was the release of the book Social Responsibilities of the Businessman by the American economist Howard Rotthmann Bowen in 1953 (University Of Iowa, 2022). In this compendium, the author deals in a structured way with the relationships of ethics and responsibility that an entrepreneur must assume in his vital role.

Although almost 70 years old, it is a book that supports CSR activities up to the present day, especially during crises, which will always be an opportunity to understand which actions, isolated or collective, of a business group or business segments had repercussions, negative ones in people's daily lives.

Although the title ends with businessmen, one should also include executive women, who, perhaps at the time, did not occupy as many positions as they do today (although still far from an ideal, inclusive number).

Since 1953, the theme has been adapting and improving, reaching other countries, modifying thoughts or at least making them more critical in the face of the need for harmonious interaction between society, organizations, and the environment in such a way that it forms several worldwide actions for this awakening, especially with the environment, which directly affects the other dimensions.

Thus, conferences, meetings, treaties, and letters to which several signatory countries and host countries were highlighted as global efforts to awaken the aforementioned critical thinking to a less compromised life through Sustainability. To highlight, the events are mentioned as shown in Table 6.1:

With these events of great repercussion, with supporters and non-supporters, with consensus and sometimes not so much, CSR has been expanding and conquering reference organizations in the business world. Large corporations began to adopt procedures in their financial statements aiming at Transparency, a concept widely applied in the 1990s, especially after the economic scandals in the US with the company Enron, for example, which led, among other actions, to the creation of the Sox Law—Law Sarbanne-Oxley by the Congress of that country in 2002.

Table 6.1 Timeline of the sustainability concept

1972—Stockholm Conference—1st UN Conference on the Human Environment
1982—Nairobi Conference, Kenya
1987—Concept of Sustainable Development. Report of the World Commission on Environment and Development—UN. It is known as the Brundtland Commission Report (named after the former Prime Minister of Norway, Gro Harlem Brundtland, who chaired the Commission above). In Brazil, the publication "Our Common Future" by the publisher of FGV—Fundação Getúlio Vargas contains the report
1992—Eco 92 or Rio 92—the UN Conference on Environment and Development in Rio de Janeiro
1997—The Kyoto Protocol, for a stricter reduction of greenhouse gas emissions, was signed in Kyoto, Japan
2002—World Summit on Sustainable Development, Johannesburg, known as Rio + 10
2012—United Nations Conference on Sustainable Development, Rio de Janeiro, known as Rio + 20
2015—Paris Agreement, Paris

In 2015, another law cited this legislation to encourage innovation in the United States for emerging growth companies for five years by applying the Sox Law (HOUSE COP, 2022).

In this context, there is a scenario of changes driven by the thinking of sustainability, responsibility toward stakeholders, the growth of thinking at Transparências, and CSR, not just as a marketing figure. In the following items, some essential definitions are resumed to highlight the topic's relevance.

Some topics, such as Ethics, Transparency, and Sustainable Development, need to be remembered better to understand the term corporate social responsibility (CSR).

Friedlan and Jain (2020) highlight how important it is for business schools to show their students a concern for social purposes due to numerous scandals involving ethical issues at the highest level of organizations, which affects social inequalities and climate change.

The concept of Ethics, although broad and very variable depending on the nationality of an organization, for example, or of society as a whole, varies depending in particular on the origin of that country—the workplace ethos. What is allowed in some will never be allowed in another, at least as long as millenary precepts of behavior prevail. The word originates from the Greek ethos, meaning "the uses and customs in force in a society and also, secondarily, individual habits" (COMPARATO, 2006).

This term, then, understands not only the customs of peoples and societies but also, in the case of CSR, the importance of observing the individual behavior of a CEO. This behavior is highly influential in an organization and its surroundings, favoring integration or disaggregation relationships—the ethical relationships supporting decision-making processes in a CSR context.

In addition, and as part of ethical actions, there is Transparency, which is currently very evident in Brazil through the Transparency Portals, whose objective is to publicize the rendering of accounts, price assessments, and allocation of resources among the public in three dimensions: Municipal, state, and federal. In this particular case represented on the website www.portaltransparência.gov.br, created by the Comptroller General of the Union in 2004 (CGU, 2022), which will be dealt with later in this chapter as an example of Social Responsibility action in an entity public.

Transparency thinking is also found in the business world, as highlighted by authors Tapscott and Ticoll (2005) and, more recently Waddock, S. et al. (2017). The concept applied to Organizations goes beyond publishing economic-financial reports but brings managers to a commitment to stakeholders, promoting competitive capacity, social respect and care for the environment, all presented to everyone, with seriousness and truth.

This understanding aligns with the concept of Sustainable Development, as being "one that meets the needs of the present without compromising the ability of future generations to also meet their own". This classic definition, currently studied in all segments, originated from the Brundtland Commission Report, published in Brazil with the title Our Common Future. In a reduced form, the expression Sustainable Development became recognized as Sustainability.

Currently, the entire planning and decision-making process in organizations needs to focus on Sustainability because a wrong decision can generate costly environmental liabilities, social commitments that are difficult to recover, and environmental impacts that are difficult to mitigate. Sustainability is no longer a concept to become the basis for all organizational processes and their internal and external reflections.

In this way and with these essential elements, Corporate Social Responsibility is established as a modern and necessary mechanism for Organizations that aim to maintain and expand their activities in the coming decades. In Brazil, since 2004, organizations can count on a Brazilian Standard for Social Responsibility—Management System, NBR ISO 16001, which aims to establish "the minimum requirements for a social responsibility management system, allowing the organization to formulate and implement a policy and objectives" (INMETRO, 2022) with some commitments as listed in Table 6.2:

It is observed that, as mentioned above, Transparency and Ethical Behavior compose the requirements of the RES standard. It even defines it as:

> Responsibility of an organization for the impacts of its decisions and activities on society and the environment through ethical and transparent behavior that:

Table 6.2 Commitment to social Responsibility

(a) Accountability and transparency;
(b) Ethical behavior;
(c) Respect for the interests of interested parties;
(d) Compliance with legal and other requirements by the organization;
(e) Respect for international norms of behavior;
(f) Respect for human rights and
(g) The promotion of sustainable development

- Contribute to sustainable development, including the health and well-being of society;
- take into account stakeholder expectations;
- comply with applicable law and be consistent with international standards of behavior and
- be integrated throughout the organization and practiced in its relationships (INMETRO, 2022).

INMETRO also highlights that the standard is "possible of integration with other management requirements and establishes criteria that apply to any type and size of the organization," especially guiding those who wish (INMETRO, 2022):

- Implement, maintain, and improve a Social Responsibility management system;
- Ensure compliance with applicable legislation with its Social Responsibility policy;
- Support stakeholder engagement;
- Demonstrate compliance with the Standard by carrying out a self-assessment and issuing a self-declaration of compliance, or by seeking confirmation of its compliance by interested parties or parties external to the organization, or by seeking certification of your liability management system by an external organization.

Before the publication of this Standard, Melo Neto and Fróes (2002, p. 84) already considered that "Social Responsibility is seen as a company's commitment to society and humanity in general and a form of accountability of its performance [...]," that is indispensable conditions for Sustainability.

After the publication of NBR ISO 16001, Ashley (2005) considered that:

> Being socially responsible implies, for the company, valuing its employees, respecting the rights of shareholders, maintaining good conduct relationships with its customers and suppliers, maintaining or supporting environmental preservation programs, complying with the legislation pertinent to its activity, collecting taxes, supporting or maintain actions aimed at reducing or eliminating social problems in the areas of health and education and providing information about its activity. In summary, the company must have the concepts of ethics and transparency as the basic principles of its conduct (ASHLEY, 2005, p. 173).

Once again, Ethics and Transparency have been essential requirements for Social Responsibility. From 2010 onward, Organizations also started to rely on ISO 26000, and in 2012, the NBR ISO 16001 was updated based on it.

The International Directive ISO 26000, unlike ISO 16001, is a standard of Guidelines (INMETRO, 2022), not certifiable, which does not make its implementation, concepts, and procedures unfeasible in Organizations worldwide. It is important to note that in Brazil, the body responsible for the standards is the Associação Brasileira de Normas Técnicas, with the acronym ABNT that accompanies the number of the Norms, it is the National Forum for Standardization, created in 1940, being a private and non-profit entity. He is a founding member of the International Organization for Standardization—ISO, of the Comisión Panamericana de Normas Técnicas—Copant, and of the Asociación Mercosur de Normalizacion—AMN (ABNT, 2022).

It is essential to highlight that about ISO 26000, Brazil, together with Sweden, was elected to direct the working group responsible for building the Standard above. The meeting that elected them was held in Stockholm, Sweden 2004, organized by ISO itself (INMETRO, 2022).

Since then, other meetings have taken place for the construction of the Standard, the first of which was held in Salvador, Bahia in March 2005, followed by others: The second meeting in Bangkok, Thailand, in September 2005; the third meeting in Lisbon, Portugal, May 2006; the fourth meeting in the city of Sydney, Australia, in January 2007; the fifth meeting in Vienna, Austria, held in November 2007; the sixth meeting in Santiago, Chile, which took place in September 2008.

The standard expected to be published in 2008 was published only in 2010.

6.4 Conceptual Framework for Implementing Corporate Social Responsibility

Considering the concepts of Ethics, Transparency, and Sustainability with a focus on Corporate Social Responsibility and verifying its importance for Organizations when the business world is rethinking its management process, this section proposes a Conceptual Framework for implementing CSR to increase sustainable competitiveness.

However, before proceeding, an important warning: All organizations, regardless of size, can implement the actions described, adapting them to their realities.

1. Prepare an organizational code of ethics—observe local customs and habits and respect the diversities and native characteristics of the city and region where the enterprise is or will be implemented. The code of ethics must meet the micro and macro corporate environment;
2. Adopt measures to disseminate the Organization's objectives and goals—publish financial reports with the necessary explanations so that all employees can understand, involving everyone in the pursuit of achieving the objectives and goals;
3. Staff training—check which skills need to be oriented and which training can be carried out in partnership with specialized bodies, through partnerships with Universities and Business Schools. Make the organizational environment more attractive to new ideas and technologies, always involving employees;
4. Recycle—maintain regular recycling of everything that can be recycled and promote the regular disposal of waste, including sanitary waste. In this case, observe if there is adequate capture. In the case of contaminated waste, apply the relevant legislation, in addition to seeking alternatives for the use of products with less environmental risk;
5. Water reuse—large organizations, composed of several roof coverings, must establish a system for capturing, adequate reserve, and reuse in production processes or cleaning shared spaces and restrooms. The smaller ones, in the

same way, especially those in regions of constant precipitation, can opt for reuse similarly. In office buildings, propose adequacy as well;
6. Integration with the surrounding community—promote constant dialogue with local society through representations of organized civil society and residents' associations by the Organization's social, environmental, and economic impact. Promoting, fostering, or enabling actions in the fields of health and education, in its way, by hiring professionals or through qualified and experienced institutions, and may even carry out public–private partnerships, building rooms and offices inadequate space and without compromising the routine activities of the company, passing the use to the public authority, duly agreed;
7. Supply Chain Management—the supply chain management process is essential so that the Organization is not surprised by suppliers who have working conditions similar to slavery or child labor, in addition to aspects of environmental impact and commitment to the environment, without forgetting tax evasion.

6.5 Cases of Corporate Social Responsibility in Public and Private Companies

As previously mentioned when dealing with Transparency, the Transparency Portal (CGU, 2022), created by the Comptroller General of the Union in 2004, aims to broadly disseminate government actions, in particular the allocation of resources.

This model is in line with Corporate Social Responsibility. ICTs—information and Communication Technologies—are available to all Organizations and should be used attractively. As presented in Sect. 3, they are a means of making financial issues, environmental actions, and social actions evident to society. The Transparency Portal is a good example of Organizations using cyberspace to publicize their actions equally.

Companies from the private and public sectors were selected to learn about CSR actions: Natura, Saint-Gobain, SABESP, and INB. Below are these organizations' main social responsibility actions and practices, which are disclosed on their websites.

Box 4.1–Natura Company–(NATURA, 2022).

Grupo Natura & Co, a cosmetics company created in São Paulo in 1969, is currently the largest Brazilian multinational. It has been publicly traded on the São Paulo stock exchange since 2004. In 1995, it created a line of non-cosmetic products called Crer para Ver, with the profits reverted to projects in education (the Instituto Natura was created in 2010 to manage these resources).

As of 2006, they stopped testing products and ingredients on animals. In 2007, they created the Carbon Neutral Program to reduce greenhouse gas emissions throughout the production chain, achieving a 1/3 reduction in emissions in 2013.

As presented in its 2020 annual report, its vision is: "By its business behavior, by the quality of the relationships it establishes and by its products and services, it will be a world-renowned brand, identified with the community of people who are committed to the building of a better world through a better relationship with themselves, with the other, with the nature of which they are a part, with the whole."

Natura is a member of international organizations and has Certifications that, according to the company itself, guarantee its sustainability actions, which are:

- B CORP: Associates are companies that combine economic growth with social and environmental promotion, a movement created in the United States in 2006. In 2019, Natura was considered the largest cosmetics company in this group (https://www.bcorporation. net/pt-br/).

Leaping Bunny Program: A global program that presents production standards without cruelty to animals, in addition to the legislation granted by Cruelty-Free International (https://crueltyfreeinternational.org/).

- UEBT-Union for Ethical Biotrade: This association's seal establishes that the organization's products concerning people, forests, and biodiversity were acquired. For Natura, this seal establishes "fair trade, conservation of Brazilian biodiversity, and relationships of trust with the community." UEBT was created as an arm of the United Nations–UN in 2007 (https://www.ethicalbiotrade.org/).

Box 4.2–Saint-Gobain–(SAINTGOBAIN, 2022).

Saint-Gobain is a French multinational that has existed for almost 360 years. It is active in the Civil Construction market and is present in Brazil through brands in this segment, such as Brasilit, Carborundum, Norton, and Quatzolit.

It is present in 70 countries and five regions of Brazil, with factories and points of sale. It has more than 167,000 employees. The company establishes principles of behavior and action for its employees.

There are Five Principles of Behavior.

1. Professional Commitment;
2. Respect for people;
3. Integrity;
4. Loyalty;
5. Solidarity.

There are Four Action Principles.

1. Respect for legality;
2. Respect for the environment;

3. Respect for health and safety at work;
4. Respect for employee rights.

The company has signed international pacts to combat climate change, including the United Nations Global Compact, which endorses care for the climate and the protection of water resources as part of the millennium goals established by the United Nations.

At COP 21, it was even an official partner of the event. It set targets to reduce its energy consumption by 15% and 20% of total CO2 emissions by 2025 and reduce industrial water waste by 80% in 10 years.

Regarding Human Resources, it signed the declaration of support for managing human rights with the UN in 2008, the 60th anniversary of the Universal Declaration of Human Rights.

Among the purposes established by the Organization, the following can be highlighted: "We conduct business respecting our principles of behavior and action and the humanist values that permeate our corporate culture. Listening, dialogue, care, solidarity, trust, and respect for difference are central points of our commitment."

Box 4.3–SABESP—Basic Sanitation Company of São Paulo—(SABESP, 2022).

Companhia de Saneamento Basico de São Paulo, identified by the acronym SABESP, is a mixed-capital corporation part of the indirect administration of the State of São Paulo. It was founded 51 years ago and is present in 370 municipalities in that state, including the capital, carrying out water distribution, capitation, and sewage treatment.

It produces, supplies, and sells reused water from sewage treatment in partnership with another company that supplies the Capuava Petrochemical Complex. It also works with non-domestic sewage and energy generation. It offers consultancy in the area of rational water use. Since 1998, it has universalized access to water where it operates.

Its environmental guidelines have established 2008 actions to care for the environment, which are.

1. Acting to prevent water pollution and solid waste management;
2. Develop people to promote the continuous improvement of products, processes, and services, aiming at environmental quality;
3. Ensuring compliance with environmental legislation and commitments made;
4. Adopt environmental criteria for supplier management;

5. Foster the development of technologies to protect, conserve, and recover the environment.

SABESP has been publishing its sustainability report for 14 years. Since 2007, it has adopted the Global Reporting Initiative—GRI guidelines to prepare the document. The company is a signatory to the United Nations Global Compact for the Environment.

Its vision is "to be a world reference in the provision of sanitation services, in a sustainable, competitive and innovative way, with a focus on the customer."

Box 4.4–Nuclear Industries of Brazil–INB–(INB, 2022)

Founded in 1988, Indústrias Nucleares do Brasil succeeded the former Nuclebrás and the group of companies that comprised it, founded in 1974. It is a public company linked to the Ministry of Mines and Energy, which monopolizes the production and commercialization of nuclear materials in Brazil.

Headquartered in the city of Rio de Janeiro with an office in the city of Fortaleza, having uranium as raw material, operating in mining, conversion, enrichment, reconversion, production of pellets, assembly of the fuel element of this main component for the production of nuclear material.

Through social communication, it maintains a direct relationship with the cities where its facilities are present, allowing visits to the units. It also carries out the Conversa com a Vizinhança Program and the Radioactivity Program.

Due to the risks of the production process, the company focuses primarily on environmental monitoring, evaluating, and preventing possible ecological impacts. This monitoring evaluates "rainwater, water courses and underground wells, soil, dust, agricultural products, and other sources."

The team of collaborators is made up of chemists, physicists, and engineers who survey the reality of the installation sites before they take place so that after the start of activities, they can carry out the monitoring, generating reports for the follow-up of inspection bodies such as the National Nuclear Energy Commission (CNEN) and the Brazilian Institute for the Environment and Renewable Natural Resources (IBAMA).

As a vision, the company declares "to be an internationally recognized company as a supplier in the uranium market, with excellence in business management, in search of self-sufficiency."

6.6 Indication of How the Conceptual Framework Will Be Incorporated into the Process of Development, Implementation, and Monitoring of the Strategy and Production Organization

Given what was exposed in the previous sections and as a way of making it possible to implement management with Social Responsibility, considering the author's studies on Ethics, Transparency, and Sustainable Development, this section proposes suggestions for the implementation of the presented Conceptual Framework:

Code of ethics: Its purpose is to guide the work of the entity and serve as a parameter for judging inappropriate attitudes in professional relationships. It has yet another purpose: determining the Organization's position vis-à-vis society, particularly the society around it.

This code also establishes the company's relations with the market, the means by which suppliers will be selected, admission and dismissal policies, and its responsibility to the environment, among other things. It is important to always understand local peculiarities and social and cultural diversity so that the company does not suffer sanctions or offend any established local principle.

Its construction can be done by promoting the participation of the internal and external community of interested parties based on another code of proven efficiency, including benchmarking.

Establish and disclose objectives and goals—essential for management that proposes transparency. This should start with the sharing of financial information, using intranets or electronic correspondence with employees. Panels can be useful as long as they are dynamic and placed in suitable and easy-to-view locations, but nothing is more effective than a direct message to the employee.

In addition to financial data, sharing objectives and goals with employees, directors, and shareholders promotes cohesion and direction of activities. Knowing where you are going is very important. It should also be noted how necessary it is for everyone to participate in the decision-making process to establish such objectives and goals and generate interaction and more commitment from all.

Staff training—Every Organization has a wealth of possibilities, and its operational and management bodies have potential that needs to be improved. Most higher education courses have difficulties finding partners to apply new techniques, adapted management models, and improved production processes. A partnership with these spaces for the development of thought is very favorable for the training of manpower from customer service to the presidency of the Organization. Everyone can learn more in the corporate environment.

Business schools are attractive partners that support staff training. From the factory entrance to the presidency's room, something changes, adjusts, and improves. There are organizations with specialized consultants for such training, but this partnership alternative must be considered, as the Organization will only sometimes have the resources to hire large consultants.

Recycle—and why not recycle? Recycling begins with evaluating all the Organization's processes. Everything must be evaluated and observed with criteria, and small alternatives will certainly be perceived that, as a whole, will represent an expressive value. There is always something to recycle, and therefore, you need to evaluate yourself.

It starts with the company and starts by encouraging employees. They are part of the process and also generate recyclable material in their personal lives. Encouraging this practice promotes a recycling network. Business does encourage personal doing.

Likewise, the Organization can encourage or participate in recycling programs with society, support educational activities, participate in sectorial events on the subject, and be a reference element in this reuse process.

Water reuse—Perhaps this is the item that will need the most investment at first, as the capture and reserve of rainwater, for example, requires special treatment and adequate equipment. It is affordable but with its own technology. It is not enough to collect the water and leave it in drums outdoors, as they can become a focus for disease vectors. Other forms of water reuse used in the production process must be evaluated.

Reuse allows the reduction of water consumption and also the concern with this natural good that has a high cost to become potable. Reuse actions make it possible to reduce the direct water consumption and, consequently, reduce costs, enabling the installation of a water treatment plant.

Integration with the surrounding community—normally, the employees of an Organization are neighbors of the enterprise, or at least they are in the same region. Some offshore businesses modify this condition. In the others, the neighborhood is formed by those directly and indirectly related to that enterprise.

This facilitates the approximation with the community, which has its leaders and representations. The Organization connects with local realities through Dialogue tools and interacts in collective, sustainable, ethical growth and respect for traditions and customs. Other items of this Conceptual Structure proposal always refer to a harmonious and dialogic relationship with the communities surrounding the Organization.

Supply Chain Management: suppliers represent an important segment in Organizations' lives. Without them, many businesses become unviable. However, the qualification and selection of suppliers should continue as a precautionary instrument for future problems.

An illegal supplier may, at any time, stop providing inputs that may be vital for the Organization. Therefore, their accreditation requires care. The customer can and must request documents that prove the existence of fact and law of a supplier so that its legality is preferred to another supplier that does not exist by right. Evaluate the volume of labor lawsuits against such companies, civil lawsuits, criminal lawsuits, whether environmental crimes were committed, etc.

In addition, the risk of hiring suppliers without any concern for social, ethical, and environmental issues can compromise the supply itself and denigrate the client's image, not to mention issues related to child labor and labor analogous to slavery.

Nowadays, the management of an Organization's supply chain must encompass these aspects as well, in addition to the discontinuity of inputs and services necessary for the business's activities.

6.7 Summary of Main Ideas and Concepts

Ethics is associated with a people's or society's customs and represents specific moral attitudes of certain countries. It is the way in which people make their decisions and interact with others.

Transparency—in the case of Organizations, it is related to how it communicate its objectives, goals, and results. Demonstrates an attitude of seriousness and commitment to stakeholders

Sustainable Development—development of current generations that does not compromise the development of future generations due to scarcity or lack of human conditions for subsistence.

Sustainability—sustainability can be understood as the word that represents sustainable development, having the same meaning as this expression.

Corporate Social Responsibility—business management that respects and promotes human diversity, respects and protects the environment, promoting Sustainability with Ethics and Transparency. That encourages education and cares for the occupational and social health of employees and society. The Organization makes Dialogue an instrument to support the decision-making process.

6.8 Questions for Reflection and Study

(1) What challenges may an Organization encounter when planning the implementation of Corporate Social Responsibility?
(2) What benefits can an Organization receive when implementing Corporate Social Responsibility?
(3) Cite three examples of actions to promote Transparency in a company.
(4) Observing the cases of Organizations presented in Sect. 4, highlight which actions are common among them. Access the websites of the Organizations to answer this question better.

6.9 Cases to Be Resolved

(a)(a)(a)(a)(a)(a) A tire manufacturing industry whose distribution takes place through different modes of transport to deliver its products. This industry has already carried out several sustainable actions in its factory, including carrying out reverse logistics. How can it mitigate the environmental impacts related to the transport of its products?

(b)(b)(b)(b)(b)(b) An executive from a large Organization receives the mission to expand the company's business in the iron mining segment. The company receives from shareholders the financial support it needs for the requested expansion. The city that has the potential to install the activity is a small municipality with an exemplary quality of life. What Corporate Social Responsibility actions should this Executive consider before making the final decision?

References

ABNT Available at https://www.abnt.org.br/institucional/sobre. Accessed Feb 2022

ABNT (2022) NBR ISO 19650-1 Versão Corrigida:2022-05-27, Publication date: 2022-05-27, https://abnt.org.br/. Accessed on March 30, 2025

Ashley, Patrícia Almeida (coordenação). Ética e Responsabilidade Social nos Negócios. 2.ed. São Paulo: Saraiva, 2005

B CORP. Available at https://www.bcorporation.net/pt-br/. Accessed Feb 2022

CGU. Controladoria Geral da União. Available at https://www.portaltransparencia.gov.br/. Accessed Feb 2022

Comparato, F. K. Ética: direito, moral e religião no mundo moderno. São Paulo: Companhia das Letras, 2006

Cruelty-Free International. Available at https://crueltyfreeinternational.org/. Accessed Feb 2022

Elkington J (2018) Sustentabilidade, há 25 anos eu criei a expressão "Triple Bottom Line". Saiba porque é hora de repensá-la. Harvard Business Review

Friedland J, Jain T (2020) Reframing the purpose of business education: crowding-in a culture of moral self-awareness. J Manag Inq 31(1):15–29. https://doi.org/10.1177/1056492620940793 (Original work published 2022)

HOUSE COP, 2022. Available at https://www.gop.gov/bill/h-r-4139-fostering-innovation-act-2015/. Accessed Feb 2022

INB. Indústrias Nucleares do Brasil. Available at https://www.inb.gov.br/. Accessed Feb 2022

INMETRO. Available at http://www.inmetro.gov.br/qualidade/responsabilidade_social/historico-iso.asp. Accessed Feb 2022

Melo Neto, F. P. de e Fróes, C. Responsabilidade Social & Cidadania Empresarial: a administração do terceiro setor. 2.ed. Rio de Janeiro: Qualitymark Ed., 2002

Natura. Available at https://www.natura.com.br/. Accessed Feb 2022

SABESP. Companhia de Saneamento Básico do Estado de São Paulo. Available at https://site.sabesp.com.br/site/interna/subHome.aspx?secaoId=3. Accessed Feb 2022

SAINT-GOBAIN. Available at https://www.saint-gobain.com.br/. Accessed Feb 2022

Sheedy E, Garcia P, Jepsen D (2020) The role of risk climate and ethical self-interest climate in predicting unethical pro-organizational behavior. J Bus Ethics 173:281–300

Stock T, Obenaus M, Kohl H (2018) Industry 4.0 as enabler for a sustainable development: an assessment of its ecological and social impacts. Manuscript Draft. Process Safety and Environmental Protection

Tapscott, Don e Ticoll, David. A Empresa Transparente: como a era da transparência revolucionará os negócios. M. Books: São Paulo, 2005

UEBT-Union for Ethical Biotrade. Available at https://www.ethicalbiotrade.org/. Accessed Feb 2022

University of Iowa. Available at https://www.uipress.uiowa.edu/books/9781609381967/the-social-responsibilities-of-the-businessman. Accessed Feb 2022

Waddock S, Andriof J, Husted B, Rahman SS (2017) Unfolding stakeholder thinking: theory, responsibility and engagement (Book). pp 1–320

Mirian Picinini Méxas PhD in Civil Engineering from UFF in the Management, Production, Quality, and Sustainable Development System Area. Master's degree in Systems and Computer Engineering from COPPE_UFRJ. MBA in Business Administration from UFF. Post-Graduation in the PIGEAD Course (Planning, Implementation, and Management of EaD) by UFF. Degree in Mathematics (Bachelor and Licentiate) at UFF. Certification in Project Management-PMP. Professional experience includes project management, business analysis, management of systems development and maintenance teams, working with multi-functional teams and contracted companies from the supplier selection process, evaluation, and acquisition to the service implementation, systems analysis, and programming. She is currently a professor of the undergraduate course in Actuarial Sciences at UFF. Professor of the Professional Master's Degree in Management Systems at UFF and the Interdisciplinary Doctorate in Sustainable Management Systems in the research line Management of Sustainable Organizations at UFF.

André Cantareli da Silva Post-Doctoral Student in Sustainable Management Systems at the School of Engineering at Universidade Federal Fluminense. Doctor in Civil Engineering from Universidade Federal Fluminense, with a concentration in Management, Production, and Environment. Master in Production and Systems Engineering at the Federal Center for Technological Education Celso Suckow da Fonseca–CEFET/RJ in the Development of Technologies Integrated to Products and Processes. He has an MBA in Total Quality Management from Universidade Federal Fluminense–UFF, a specialization in Planning, Implementation, and Management of Distance Learning (in progress) from Universidade Federal Fluminense–UFF, a degree in Accounting from Universidade Salgado de Oliveira, an extension in Balance Sheet Analysis from Getúlio Vargas Foundation–FGV. Graduated in Ceramic Process Technology, by SENAI/SP Mário Amato. He is a full professor at Fluminense Federal University, coordinating the Accounting Sciences course and advising Professor of the Graduate Program of the National Public Administration Program.

Chapter 7
ESG Challenges in the Social Context of the Animal Protein-Based Food Sector: The Brazilian Case

Juliana Damaris Candido De Lima, Renan Carriço Payer, Samuel Martins Drei, Lizzie Bessa Risicato, Ruben Huamanchumo Gutierrez, Lidia Angulo-Meza, Gilson Brito Alves Lima, Lincoln Campelo Dias, and Pâmela De Carvalho Marques Silva

Abstract In recent years, companies have recognized the benefits of adopting a proactive approach to the environment. As well as improving efficiency and gaining competitive advantage, they can also reduce legal compliance costs. Environmental management is the system used by companies to minimize or eliminate the negative impacts of their activities on the environment. However, many companies believe that taking measures to protect the environment will increase their production costs. However, some companies have shown that it is possible to make a profit and protect the environment at the same time. To do this, companies need to be creative and have

The original version of the chapter has been revised. A correction to this chapter can be found at https://doi.org/10.1007/978-981-96-1725-8_13

J. D. C. De Lima · R. C. Payer · L. B. Risicato · R. H. Gutierrez · L. Angulo-Meza · G. B. A. Lima · P. De Carvalho Marques Silva
Production Engineering, Federal Fluminense University, Niterói, Brasil
e-mail: julianadamaris@id.uff.br

R. C. Payer
e-mail: rpayer@id.uff.br

L. B. Risicato
e-mail: lizziebr@id.uff.br

R. H. Gutierrez
e-mail: rh_gutierrez@id.uff.br

L. Angulo-Meza
e-mail: lidiaangulomeza@id.uff.br

G. B. A. Lima
e-mail: glima@id.uff.br

P. De Carvalho Marques Silva
e-mail: pamelacarvalho@id.uff.br

M. J. Meiriño and O. L. G. Quelhas (eds.), *Critical Factors Associated with Environmental, Social and Governance Management in Organizations*, Accounting, Finance, Sustainability, Governance & Fraud: Theory and Application, https://doi.org/10.1007/978-981-96-1725-8_7

internal conditions that allow them to turn environmental restrictions and threats into business opportunities. Over the years, companies have adopted environmental management systems to improve their relationship with the environment and gain the credibility of society. This allows them to stand out from those that only fulfill legal requirements and generates a competitive advantage in the market. The concept of sustainable development is one that meets the needs of the present without compromising the ability to meet the needs of future generations. The Triple Bottom Line approach, which evaluates a company's economic, social, and environmental performance, is widely discussed in the literature. Corporate sustainability is linked to these three aspects: economic prosperity, environmental quality, and social justice. Companies that manage to turn sustainability into a business opportunity gain a competitive advantage and are recognized by stakeholders. The Brazilian animal protein industry has expanded internationally, acquiring processed food brands and serving as an example for other Brazilian companies that want to internationalize. Investing in sustainability in the animal protein industry brings financial returns and also benefits society. Furthermore, promoting improvements with a focus on sustainability increases productivity and product quality. The food and beverage industry are the largest in the country, and concern about animal welfare and the environmental impacts of livestock production has become an important issue in the management of animal protein companies. Meatpacking plants, which emerged in the 1920s, are responsible for producing fresh meat and other related products. The complexity of the production chain has made it necessary to seek quality and differentiation for products. Sustainability is a relevant topic in business management, as it contributes to reputation and brand protection. The Global Reporting Initiative provides standards for voluntary environmental and social performance reports, helping companies and organizations to take responsibility for their impacts and communicate globally. Sustainability assessment and decisions for a sustainable future must be made in a structured, transparent and reliable way. The Technique of Optimal Solution Similarity (TOPSIS) is one of the tools that can be used to assess the sustainability of animal protein companies. This research aims to develop sustainability analyses based on the Triple Bottom Line and TOPSIS for the management of these companies.

Keywords ESG · Triple bottom line · TOPSIS · Sustainability · Competitiveness · Animal protein

---

S. M. Drei (✉)
Production and Manufacturig Engineering, Federal Fluminense University, Niterói, Brasil
e-mail: samueldrei@id.uff.br

L. C. Dias
Engineering Systems Management, Federal Fluminense University, Niterói, Brasil
e-mail: lincolndias@id.uff.br

7.1 Introduction

Over the last few decades, different organizations have realized the benefits that can be obtained from adopting a proactive stance in their relationships with the environment, as well as improving their efficiency, gaining competitive advantage, and reducing compliance costs through actions aimed at legal compliance–meeting the requirements of the sector's legislation (Torres et al. 2022).

Environmental management is the system that includes organizational structure, planning activities, responsibilities, practices, procedures, processes and resources for developing, implementing, achieving, critically analyzing and maintaining environmental policy. It is what the company does to minimize or eliminate the negative effects of its activities on the environment. The first doubt that arises when considering the environmental issue from a business point of view is the economic aspect, because with any measure that is taken in relation to the environmental variable, the idea arises those expenses will increase and, consequently, there will be an increase in the costs of the production process (Goyanes et al. 2022).

However, some companies have shown that it is possible to make money and help protect the environment, even if they are not organizations operating in the so-called 'green market'. To do so, these companies must have a certain amount of creativity and the internal conditions to turn environmental restrictions and threats into business opportunities (Martins et al. 2022).

Over the years since the 1940s, companies have been adopting management systems to mitigate and improve the communication of their positive relationship with the environment, thus gaining credibility from society. This process of legitimization allows organizations to stand out from those that only comply with environmental legislation, which creates a competitive edge in the market.

The concept of sustainable development has an interdisciplinary and multidisciplinary relationship with different assumptions about the basic relationship between society and nature. These include the definition by the World Commission on Environment and Development, which states that sustainable development is development that meets the needs of the present without compromising the ability to meet the needs of future generations (Torres et al. 2022).

Based on the concept of sustainable development, methods have been developed and discussed in the literature, such as the Triple Bottom Line approach. The Triple Bottom Line concept states that the ultimate success or health of a company can and should be measured not only by traditional financial results, but also by its social and environmental performance. In other words, the economic, social, and environmental dimensions must be taken into account (Caiado et al. 2022).

The concept of corporate sustainability is based on the Triple Bottom Line, which focuses on three aspects: economic prosperity, environmental quality, and social justice. Contemporary environmental issues are increasingly linked to companies' strategies to turn sustainability into a business opportunity, which means gaining a competitive advantage and legitimizing themselves in the eyes of stakeholders.

The recent international expansion of Brazilian animal protein companies and the relevance they have achieved internationally through the acquisition of important processed food brands can serve as a success story for other Brazilian companies wishing to internationalize.

In this scenario, as one of the largest producers in the world, Brazilian animal protein companies are committed to prioritizing economic, social, and environmental factors, earning international recognition as a benchmark in sustainability (Bressanelli et al. 2019).

There is no doubt that investing in sustainability actions in the animal protein industry pays off. There are even more important gains. Sustainability in the animal protein industry also involves social work, which directly implies the development of society as a whole. Furthermore, promoting improvements with a focus on sustainability is also a way of investing in an area that effectively increases productivity and product quality.

The food and beverage industry are the largest in the country and nowadays the discussion about the sustainability of companies in the animal protein sector is a pertinent topic due to society's growing concern about animal welfare and the environmental impacts caused by livestock production (Caiado et al. 2019).

The first meatpacking plants originated in the 1920s, with the creation of multinational companies that focused on the wholesale and export of meat. From the 1960s onward, the protein market intensified and there was a huge increase in production of pork, beef, lamb, and poultry. In terms of exports and imports, beef and poultry were more in demand than sheep and pork. In the 1980s, developed countries found their animal protein markets saturated and began exporting heavily. This situation, together with the growing supply from some developing countries, resulted in a highly competitive international trade (Carvalho et al. 2019).

Today, the meatpacking industry is responsible for the production of fresh meat and other products such as leather, offal, and processed meat. According to 2020 data from the Brazilian Association of Export Industries (ABIEC), approximately 87.2% of the volume of meat exported from Brazil comes from fresh meat. There are approximately 1000 slaughterhouses in operation in Brazil. A large proportion of these meatpacking plants belong to the four major groups in the sector. The high complexity of the animal protein chain, together with growing consumer demand, has transformed commodity meat into more refined, branded cuts with greater added value. Companies are therefore increasingly looking for quality and differentiation in their products (Cattani et al. 2009).

The issue of sustainability is constantly in the spotlight and is a pertinent topic in business management. In view of this, the most relevant incentive for companies to start reporting their sustainability indicators is reputation management and brand protection. The Global Reporting Initiative (GRI) is an independent international organization that provides standard reports under the name GRI and is the best-known framework for voluntary reporting of environmental and social performance for both national companies and other international organizations. It supports companies and other organizations to take responsibility for their impacts through a common global language of communication.

Due to the complexity of sustainability assessment and the need to provide direction for a sustainable future, decisions must be made in a structured, transparent and reliable way, and multi-criteria decision analysis can contribute to this goal. The Technique of Optimal Solution Similarity (TOPSIS) is one of the multi-criteria decision analysis methods that can be used to assist sustainability analysis solutions. This method was first developed to solve a multi-attribute decision-making problem.

The food and beverage industry are the largest in the country and nowadays the discussion about the sustainability of companies in the animal protein sector is a pertinent topic due to society's growing concern about animal welfare and the environmental impacts caused by livestock production. With this in mind, this research aims to develop sustainability analyses using the Triple Bottom Line in light of the Technique of Similarity by Ideal Solution (TOPSIS) for the management of animal protein companies.

7.2 Role of the Chapter in the Structure of the Book

In the last decades, different organizations have realized the benefits that can be obtained from the adoption of a proactive positioning in their relationships with the environment, in addition to improving their efficiency, gaining competitive advantage, and reducing compliance costs through actions aimed at legal compliance—meet industry legislation requirements (Darnall 2003).

In fact, several companies have demonstrated that it is possible to obtain economic gains and collaborate to reduce the impact on the environment, in addition to promoting significant measures of a social nature. Therefore, these companies must have a certain dose of creativity and maturity in environmental management practices, planning to involve or engage stakeholders in their projects. The purpose is the synergy between the organization's objectives and those of society and the environment and shareholders, transforming restrictions and expectations of environmental activists into business opportunities (Barbieri 2016; Donaire 1999).

Over the years, starting in the 1940s, companies have adopted management systems in order to mitigate risks and improve social and environmental performance. This legitimation process makes it possible for organizations to stand out from those that only meet their obligations regarding environmental legislation, which generates a competitive differential in the market.

The concept of sustainable development, in turn, presents an interdisciplinary and multidisciplinary relationship in relation to different assumptions about the relationship between society and nature (Elliott 2013). Among them, the definition of the World Commission on Environment and Development (WCED 1987) which argues that sustainable development is "that which meets the needs of the present without compromising the ability to meet the needs of future generations".

Based on the concept of sustainable development, methods are developed and discussed in the literature, such as the Triple Bottom Line approach. The Triple Bottom Line concept is that the ultimate success or health of a company can and should be measured not only by traditional financial results, but also by its social and environmental performance (Norman and MacDonald 2004). That is, the economic, social and environmental dimensions must be considered.

The concept of corporate sustainability based on the Triple Bottom Line focuses on three aspects, which are: economic prosperity, environmental quality and social justice (Elkington 2011). So that contemporary environmental issues are increasingly connected with the strategies of companies to transform sustainability into a business opportunity, which means obtaining a competitive advantage and legitimizing itself before stakeholders.

The recent international expansion of Brazilian companies producing animal protein and the relevance they have gained internationally, through the acquisition of important brands of processed foods, can serve as a success case for other Brazilian companies that want to internationalize (Siqueira 2016).

In this scenario, as they are among the largest producers in the world, Brazilian animal protein companies are committed to prioritizing economic, social and environmental factors, acquiring international recognition as a reference in sustainability.

Undoubtedly, investing in sustainability actions in the animal protein industry has a certain return, with even more important gains. Sustainability in the animal protein industry also involves social work, which directly implies the development of society as a whole. In addition, promoting improvements focused on sustainability is also a way of investing in an area that effectively increases productivity and product quality.

The food and beverage industry are the largest in the country and currently the discussion about the sustainability of companies in the animal protein sector is a relevant topic due to the growing concern of society in relation to animal welfare and the environmental impacts caused by production livestock.

The first slaughterhouses had their origins in the 1920s, with the creation of multinational companies that focused on wholesale marketing and meat export. From the 1960s, the protein market intensified and there was a large increase in production of pork, beef, sheep, and poultry. In terms of exports and imports, beef and poultry had a stronger demand compared to sheep and pork (GIRAC 1987). In the 1980s, developed countries were left with saturated animal protein markets and began to export in weight (Mueller 1996). This situation, together with the increasing supply from some developing countries, has resulted in highly competitive international trade.

Currently, the meatpacking segment is responsible for the production of fresh meat and other products, such as leather, offal, and processed meat. According to 2020 data from the Brazilian Association of Exporting Industries (ABIEC), approximately 87.2% of the volume of meat exported from Brazil comes from fresh meat (ABIEC 2021). In Brazil, there are approximately 1000 refrigerators in operation. Among these slaughterhouses, a large part belongs to four major groups in the sector (Ribeiro 2020). The high complexity of the animal protein chain, combined with the growing demand from consumers, has transformed the commodity meat into more refined cuts, with a brand and greater added value (Pascoal 2011). So, companies are increasingly looking for quality and differentiation in their products.

The theme of sustainability is constantly highlighted and is a relevant topic in business management. In view of this, the most relevant stimulus for companies to start reporting their sustainability indicators is reputation management and brand protection (Brown, Jong and Levy 2009). The Global Reporting Initiative (GRI) is an independent international organization that provides standard reports under the name of GRI, being the best-known framework for voluntary environmental and social performance reporting for both national companies and other international organizations. There is support for companies and other organizations to take responsibility for their impacts from a common global language of communication.

Due to the complexity of sustainability assessment and the need to provide a direction for a sustainable future, decisions must be made in a structured, transparent and reliable way, and multi-criteria decision analysis can contribute to this objective (Cinelli, Coles, and Kirwan 2014). The Ideal Solution Similarity Technique (TOPSIS) is one of the multi-criteria decision analysis methods that can be used to assist sustainability analysis solutions. This method was developed for the first time to solve a multi-attribute decision-making problem (Hwang and Yoon 1981).

The food and beverage industry are the largest in the country (ABIA 2021) and currently the discussion about the sustainability of companies in the animal protein sector is a relevant topic due to the growing concern of society in relation to animal welfare and the environmental impacts caused by production livestock. Thus, this research aims to develop sustainability analyzes through the Triple Bottom Line in light of the Similarity Technique by Ideal Solution (TOPSIS) for the management of animal protein companies.

Brazil should produce 10.4 million tons of meat, while the forecast for world production is 61.5 million tons. The United States Department of Agriculture (USDA) recently released the global beef production outlook. Brazil will hit the mark of 10.4 million tons of animal protein this year, the second largest volume in the world, just behind the United States. The world should produce 61.5 million tons of beef in 2021, growing 1% compared to 2020. According to projections, the United States remains the main world producer with 20% of the global volume produced, followed by Brazil with 16 0.8% and the European Union appearing in third place with 12.5%.

7.3 Theme Development History

Concern for the environment has always been an agenda in modern society, however, its prominence came only from the mid-70 s, when the world population began to observe severe changes in the quality of air, water, and soil, due to the strong and constant environmental degradation resulting from the contributions of wild and insatiable capitalism, brought about by the need for consumption. With the end of the 1960s, it became clear that economic growth and the process of predatory industrialization were bringing disastrous results to the planet (Milaré 2014).

Therefore, the international community, through the United Nations (UN), proposed a conference to address issues related to the necessary preservation and protection of the environment and, in June 1972, in Stockholm, held the United Nations Conference on the Human Environment, with the participation of 113 countries, 250 non-governmental organizations and UN bodies. Despite this, the concept of sustainable development was only properly discussed in the Brundtland Report, in 1987, prepared by the World Commission on Environment and Development, created in 1983 by the United Nations Assembly (Almeida e Araújo 2013).

From then on, as expected, the concept of sustainable development gained prominence on the international stage (Costa 2013), reaching the point of advancing and unfolding in the idea of sustainability, and both concepts were created with the same objective of allowing the development of peoples and, at the same time, preserving the environment and the quality of life of the present and future generations (Gomes 2018).

Therefore, sustainability, starting from an evolution of sustainable development, has been slowly, over the years, being treated and studied in the national scenario, so that this idea starts to consider the requirement of respect for present and future generations (Mafra 2015).

Therefore, the concept of sustainability must permeate each and every activity, with the objective of allowing greater environmental, social, economic, and ethical balance, which together will allow the development of the legal-political dimension (Gomes e Ferreira 2018), data such as the Sustainable Development Goals (SDGs). This chapter deals specifically with the Environmental, Economic, and Social objectives of sustainability, given that they are the precursors of the criteria used.

The dimensions of sustainability must be developed and worked on together, since the deterioration of the planet's resources is unsustainable, as well as poverty, as well as social exclusion, injustice, oppression, slavery, and cultural and economic domination (Mafra 2015). Thus, it is consensual that there must be interrelationships between dimensions and that indicators that are restricted to only one dimension do not reflect sustainability (Froehlich 2014).

In terms of the Environmental dimension of sustainability, the idea that a balanced environment must be properly preserved and protected is non-negotiable, providing a healthy quality of life for present and future generations, at the risk of nature no longer supporting human life in the Earth (Amado 2016).

Some of the answers on the solution of problems facing the environment can be found with the help of other pillars, such as Social, Economic, Ethical, and among others. In this sense, full life on the planet depends on today's choices for the emergence of future generations' lives, since human needs are limitless (the result of exaggerated consumerism, created by savage capitalism), but natural environmental resources do not (Freitas 2016).

Furthermore, it is important to highlight that the Environmental dimension of sustainability presupposes a fair measure of consumption toward the environment, imposing only a paradigm shift, taking from nature only what is sufficient to achieve the objective of developing as humanity, without excesses. Therefore, the ideal of sustainability needs to be an inescapable premise of human daily life, since nature has its own natural forms of balance, but human emissions have taken the world out of this balance (Martins et al. 2022).

Therefore, the SDGs focused on the environment, such as guaranteeing a healthy life and promoting well-being for all (SDG-3), guaranteeing access to water and sanitation for all (SDG-6), guaranteeing access to clean energy (SDG-7), ensure decent work and sustainable economic growth (SDG-8), ensure sustainable consumption and production (SDG-12), adopt measures to combat climate change and its effects (SDG-13) and protect life on earth [40] (ODS-15) stand out in this study, with the aim of generating criteria aimed at the environmental scenario of sustainability.

As it is not possible to continue this unbridled model of economic growth, as exemplified above, it is necessary to adopt sustainable development to meet present demands, without compromising the right to a balanced environment to the healthy quality of life of future generations, consequently, the healthy environment and the right to peace are extensions of the right to life (Torres et al. 2022).

What is evident in the Economic dimension of sustainability is that the economy needs to be properly balanced, to the point of allowing economic growth in the long term, durable, serious, and committed to human life and the planet itself, in this way the economy cannot be separate from the mediation of consequences, long term (Caiado et al. 2022).

Therefore, the Economic SDGs, such as ending poverty and hunger in all its forms (SDG-1), encouraging sustainable agriculture (SDG-2), ensuring healthy lives, and promoting well-being for all (SDG-3), guaranteeing access to clean energy (SDG-7), guaranteeing decent work and sustainable economic growth (SDG-8), reducing inequalities in the country (SDG-10) and adopting measures to combat climate change. Climate and its effects (SDG-13) [43] guide the criteria of economic sustainability.

Finally, the social dimension of sustainability emphasizes a necessary and indispensable concern for the human being and his quality of well-being, as there is an intimate relationship between the quality of human life and the quality of the environment since they are concepts inseparable (Gomes 2018).

In the Social conception of sustainability, there is the possibility of a development based on personal social development, promoting the human being and the whole community, so that everyone can have their rights to education, housing, work, and health properly guaranteed (Amado 2016).

Therefore, in the social aspect, the SDGs addressed are aimed at ensuring a healthy life and promoting well-being for all (SDG-3) and guaranteeing decent work and sustainable economic growth (SDG-8) and ensuring sustainable consumption and production modalities. (SDG-12) are the guiding principles of the social criteria.

7.4 Presentation of the Conceptual Framework

The choice of indicators was based on the PRISMA recommendation, Preferred Reporting Items for Systematic reviews and Meta-Analyses. A checklist of 27 items recommended in systematic reviews, a drafting guide, and explanations that provide additional guidance for each item in addition to examples, have been published (Campos et al. 2020). In 2020, we identified original work items that used to be incompletely reported by reviewing 60 documents that provide guidance for reporting systematic reviews (Page et al. 2021).

Due to the characteristics of this study, some adaptations were necessary. The following analysis criteria were selected based on the PRISMA recommendation (Costa and Lima 2020):

Identification: Bibliographic search in the databases, in May 2021, the search criteria are listed in Table 7.1.

1. **Selection**: By reading the titles and abstracts of the publications obtained in the researched databases, they were selected to address the sustainability indicators that best reflect the state of organizations, through the TOPSIS methodology.
2. **Eligibility**: In this phase, a complete reading of the selected articles was carried out, considering only those that focused on the evaluation of global performance in the GRI objectives. The criteria used to exclude the articles were:
   - Did not discuss the GRI sustainability indicators.
   - Had their focus on a specific sector.
   - were not available for consultation.
3. **Inclusion**: The articles that discussed sustainability indicators, their meaning and importance were selected to define the indicators to be selected in the work. Figure 7.1 illustrates the process.

The research related to the state-of-the-art literature on GRI indicators, which initially located 1968 works, was refined according to the steps shown and, finally, we arrived at the 4 selected articles (Table 7.2) in which the most significant indicators were located.

It is noteworthy that of the indicators chosen in the articles resulting from the research, those that were available in the periods for both companies were verified in the reports of the analyzed companies. Details of each article are available in the presentation in the following section.

Table 7.1 Data from the bibliographic research

Search engine	Searched words	Language	Time course
Scopus e science direct	TOPSIS and "Sustainability Indicator"	Portuguese or english	De 2019 a 2021

Source Authors

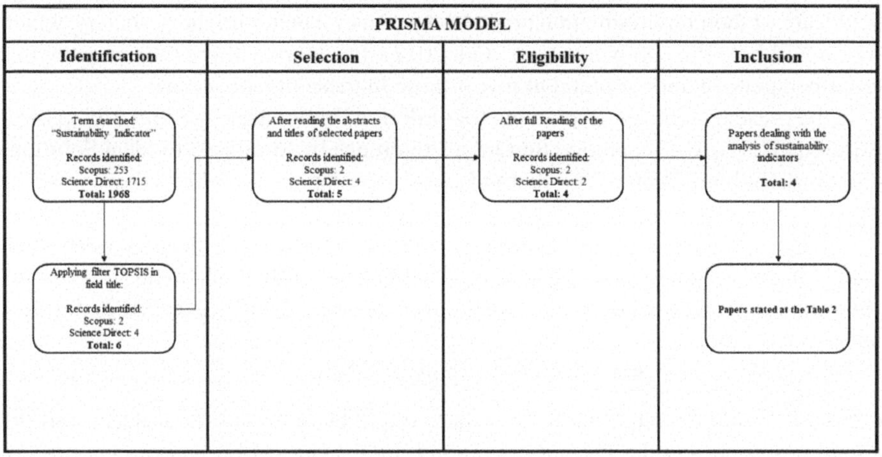

Fig. 7.1 Literature review flowchart. *Source* Authors

Table 7.2 List of selected works

Authors	Title	Year	Journal name
(Ozkaya and Erdin 2020)	Evaluation of smart and sustainable cities through a hybrid MCDM approach based on ANP and TOPSIS technique	2020	Heliyon
(Marzouk and Sabbah 2021)	AHP-TOPSIS social sustainability approach for selecting supplier in construction supply chain	2021	Cleaner Environmental Systems
(Soltani 2020)	An integrated framework using VSM, AHP and TOPSIS for simplifying the sustainability improvement process in a complex manufacturing process	2020	Journal of Engineering, Design and Technology
(Ahmed et al. 2019)	Selection of Sustainable Supplementary Concrete Materials Using OSM-AHP-TOPSIS Approach	2019	Advances in Materials Science and Engineering

Source Authors

The Global Reporting Initiative (GRI) is an independent international organization, founded in 1997 based in Amsterdam, which aims to help companies and other organizations take responsibility for the impacts caused by their activities, helping to create a sustainable future. It is responsible for issuing the most widely used standard globally for reporting sustainability (Global Reporting Initiative 2021).

In 2014, the fourth generation of sustainability reporting guidelines, commonly known as GRI-4, was launched. The use of these standards is optional, however, due to the growing interest of society in sustainable practices, more and more organizations have provided reports of their sustainability practices in accordance with the standards established in their standards. Companies can increase their degree of disclosure of their environmental practices to convey a better image to society, which in turn will help the company (Vogt et al. 2017). Investors will see the company in a better perspective and will tend to invest more in these organizations.

In the present study, the Technique for Ordering Preferences by Similarity to Ideal Solution (TOPSIS—Technique for Order Preference by Similarity to Ideal Solution) was used (Soltani 2020). The method aims to compare the alternatives with the positive ideal solution (PIS—Positive Ideal Solution).

For the application of the ranking model 21 significant indicators were used, framed in the 3 dimensions of the Triple Botton Line (TBL), which are: Economic, Environmental and Social, of two Brazilian companies (EMP1 and EMP2) during 2 consecutive years.

We will describe below the steps to apply TOPSIS.

- **Step 1**: Build decision matrix B, where the lines represent the criteria, and the columns represent the alternatives. Therefore, the values of $b_{i,j}$ represent the score of alternative A in criterion C, according to Eq. 7.1.

$$B = \left[b_{C,A}\right]_{21\times 4} \begin{bmatrix} A_1C_1b_{1,1}A_2A_3b_{1,2}b_{1,3} \\ \ldots A_4\ldots b_{1,4}C_2b_{2,1}C_3b_{3,1} \\ b_{2,2}b_{2,3}b_{3,2}b_{3,3}\ldots b_{2,4}\ldots b_{3,4} \\ \vdots C_{21}b_{21,,1} \\ \vdots b_{21,2}b_{21,3}\vdots\ldots b_{21,4} \end{bmatrix} W_iw_1w_2w_3\vdots w_{21} \tag{7.1}$$

- **Step 2**: Weight and normalize the decision matrix, through Eq. 7.2, where the values of w_i are the values of the weights assigned to each criterion.

$$n_{i,j} = \frac{w_i \times b_{i,j}}{\sqrt{\sum_{i=1}^{21} w_i \times b_{i,j}^2}} \tag{7.2}$$

- **Step 3**: Using Eqs. 7.3 and 7.4, respectively, calculate the positive (PIS, A +) and negative (NIS, A−) ideal solutions. Where $n_i +$ represents the best score achieved and n_i- the worst score achieved in the alternatives in the criterion to which it corresponds.

$$A^+ = \left\{\text{MAX}in_{i,j}|i = 1, 2, 3, \ldots, 21\right\} = \left\{n_1^+, \ldots, n_j^+, \ldots, n_i^+\right\} \tag{7.3}$$

$$A^- = \left\{\text{MIN}in_{i,j}|i = 1, 2, 3, \ldots, 21\right\} = \left\{n_1^-, \ldots, n_j^-, \ldots, n_i^-\right\} \tag{7.4}$$

- **Step 4**: Determine the distance (Di +) between the elements of matrix B, after normalizing and weighting, and the positive ideal solution through Eq. 7.5. Analogously, we must determine the distance (Di−) of the elements of matrix B, after normalizing and weighted, and the negative ideal solution through Eq. 7.6.

$$D_i^+ = \sqrt{\sum_{i=1}^{n} \left(n_{i,j} - n_i^+\right)^2} \tag{7.5}$$

$$D_i^- = \sqrt{\sum_{i=1}^{n} \left(n_{i,j} - n_i^+\right)^2} \tag{7.6}$$

- **Step 6**: Finally, through Eq. 7.7, the approximation coefficient (CCi) of each alternative is calculated and the ordering is performed in descending order. The closer to 1 the value of CCi, the better the performance of the alternative.

$$\mathrm{CC}_i = \frac{D_i^-}{D_i^+ + D_i^-} \tag{7.7}$$

The Brazilian food and beverage industry are the largest in the country: it represents 10.6% of the Brazilian GDP and generates 1.68 million formal and direct jobs (Ahmed et al. 2019).

The vegetarian and vegan food markets represent a niche market, which has been growing worldwide, leveraged by the growing concern of consumers with the impact of food on their health, animal welfare and the environmental impact of production systems, especially livestock production (Révillion et al. 2020), thus representing a challenge for the consolidation of sustainable practices in these companies.

Therefore, the relationship between food and sustainability, despite being an old topic, is still a current and relevant discussion. Therefore, the case study of this research was developed from information reported in the sustainability reports of two companies in the food industry, more precisely in animal protein. The first being a large company and one of the largest food companies in the world. The second is a multinational company that has a diversified product portfolio.

Data and indicators were collected from the Annual Sustainability Report of EMP1 and EMP2, made available on the website of the respective companies, on May 6, 2021, data related to the pillars of sustainability presented in the Triple Bottom Line approach: economic sustainability, social and environmental that were published by the aforementioned companies in the GRI for the years 2018 and 2019. In all, the research considered with 21 criteria, 11 of the Environmental dimensions, 6 of the economic dimensions, and 4 criteria of the social dimension. In the selection of these criteria, attention was paid to the similarity of the units of the chosen criteria and their homogeneity, to compare information that is in fact analogous.

7.5 Applications with the Presentation of Similar Cases in Different Contexts

Box 4.1: Case Study of the Cehab Channel

The research by Campos et al. (2020) addresses the case study of the Cehab open channel, located in the center of the urbanized area of the municipality of Itaperuna-RJ and the problems of the river water course that causes some floods. In the methodology of the work, there is a description of the field of study with images of maps and the city, the conceptual framework idealized in the work, the diagnosis of the watercourse and riverside conditions. The multi-criteria analysis method is introduced, citing other methods and their direct and indirect relationships with the TOPSIS Method. There is a description of the steps of the TOPSIS Method, from the elaboration of the decision matrix to the classification of alternatives according to global performance, demonstrating that the highest global performance was for alternative T7 (geogrids and geocells), both for the Reach1, as for Reach2. This study sought to help the decision-making process associated with the sustainable management of water courses in urban areas, selecting the best intervention techniques to be implemented. The open channel of Cehab, located in the urban area of the medium-sized municipality of Itaperuna (Northwest region of the state of Rio de Janeiro-Brazil), was used as a case study. This watercourse is often subject to flooding (associated with urbanization), affecting the local population, which urgently requires engineering intervention. In this sense, a primary diagnosis of the conditions of the watercourses and riversides made it possible to develop a set of traditional and sustainable techniques to respond to local intervention requests. These techniques were then analyzed based on sustainability indicators that include environmental, social, and economic dimensions. Thus, we used the Multi-criteria Decision Method (MCDM) called Order Preference Technique by Similarity to the Ideal Solution (TOPSIS), which is based on the evaluation of benefits and costs, to analyze the possible intervention techniques in water courses of a sustainable point of view and classify them in order to indicate which technique would be most suitable for each section of the watercourse studied. The multi-criteria analysis for Reach 1 (which presents both banks in natural condition but with advanced erosion) showed an interventionist technique (T7) as the most suitable to be implemented, followed by 7 other techniques. For Reach 2 (which has both banks covered by a conventional concrete wall), the results also showed the same T7 technique as the most indicated, however, followed by 7 other techniques. This evidence the variety of results according to the current conditions of the studied watercourse

Finally, this study showed that the multi-criteria methodology developed here can be used to guide sustainable investments, helping decision makers, based on sustainability indices. Furthermore, this methodology has a relatively quick implementation and can be later applied to other watercourses in urbanized areas, especially in developing countries, performing a primary analysis to define the appropriate techniques that could be applied to watercourse interventions.

Box 4.2: Case Study in the Energy Context

The study by Caiado, Quelhas, Nascimento, Anholon, e Leal (2019) contextualizes the concept of sustainable development, disseminated through the Brundtland report (1987), and relates it to some methods in the literature, e.g., Triple Bottom Line (TBL) and Sustainable Return on Investment (SROI). Although the text addresses the concept of the Global Information Initiative (GRI), it could have gone a little deeper into the explanation and discussion of this term, since it has a significant importance in the homogenization of data and analyzes with different variables. Additionally, although it has no direct influence on the calculations, it would be interesting to raise the Sustainable Development Goals (SDGs), related to each GRI, mainly because it is research in the energy company area, to encourage discussion in possible future works in in relation to ODS-7, information that is made available in the reports of the companies surveyed, and that would be important. The support of the use of the TOPSIS method is guaranteed, both by citing other methods of multi-criteria analysis, e.g., SMART, AHP, ELECTRE, PROMETHEE, TODIM, UTA-CR and VIKOR, and through examples in the literature, supporting the advantages of using of the method. The methodology presents a robustness in the use of the method, by demonstrating the stages of its use and the mathematical concept involved

The analysis of similarity by ideal solution in the 7 energy companies studied, demonstrated the criteria to evaluate sustainability, the evaluation matrix for the years 2013 and 2014, the ideal solution (PIS) and anti-ideal (NIS), the Distances Euclidean, the R ratio, and the ranking order, in the two years Company 1 presented first place in the ranking of hierarchies. The discussion of the results is supported by laws, provisional measures, EBITDA, the National Interconnected System (SIN), and the DJSI World index, providing information for benchmarking and suggesting future studies with clear and adequate sustainability issues.

Box 4.3: Study Focused on Industrialization and Social Inequality

The study by Paz, Caiado, Quelhas, Gavião, e Lima (2021) was applied in the state of Maranhão, considering that among the 7 states with the worst poverty rate, it has the lowest per capita income and the second largest population. The increasing impact of industrialization and changes in society's consumption behavior are associated with climate change, pollution, and increased social inequality. This fact has increased the concern with environmental awareness, encouraging organizations such as the UN to promote conferences and agreements such as Millennium Development Goals (MDGs) or the Sustainable Development Goals (SDGs) that seek to promote shared prosperity and management of natural resources, among others. TOPSIS was the method chosen among some available (AHP, ELECTRE, DEA, TOPSIS AND PROMETHEE), due to its simplicity and ability to evaluate different alternatives with different criteria, based on the shortest distance from the ideal point and the greater than the anti-ideal point. The performance indicators were identified in 17 scientific documents published between 2010 and 2018. We arrived at 33 indicators divided into 3 dimensions. The decision matrix, in the case under study, has 217 lines (alternatives–municipalities) and 31 columns (indicators–criteria). 31 indicators were used due to the unavailability of CO2 emissions and internet access in all evaluated municipalities.

7.6 Indication of How the Conceptual Structure Will Be Incorporated into the Process of Development, Implementation, and Monitoring of the Strategy and the Production Organization

The research was conducted in the two Brazilian animal protein companies, having as parameters their similarities in relation to the sector that fit the economy, as well as their end-activities. After verifying the last sustainability report of the respective companies, with the data related to the years 2018 and 2019, the GRI's stratification was performed, considering the *Triple Bottom Line model*, according to the dimensions: environmental, economic, and social. The research had 11 criteria of the Environmental dimension, 6 of the economic dimensions, and 4 criteria of the social dimension. Table 7.3 presents the criteria used to assess sustainability.

In addition to the similarity of the units of the criteria, e.g., m^3 (cubic meter), $_{tCO2}$ (tons of CO2 equivalent), GJ (Gigajoule), t (ton), R$ MM (in millions of reais), and % (percentage), was concerned with the correspondence of the GRI's chosen for the attribution of the criteria. The GRI's corresponding to the dimensions applied in the two companies studied were:

Table 7.3 Criteria for assessing sustainability

Dimension	Standard	Measurements	Units
Environmental	ENV1	Surface water	m^3
	ENV2	Ground water	m^3
	ENV3	Rainwater directly collected and stored by the organization	m^3
	ENV4	Abastecimento püblico e terceiros	m^3
	ENV5	GHG hnissions-scope 1	tC02
	ENV6	GHG anis sions-scope 2	tC02
	ENV7	GHG Enissions-scope 3	tC02
	ENV8	Energy non-renewable sources	GJ
	ENV9	Energy renewable sources	GJ
	ENV10	Hazardous waste	t
	ENV11	Non-hazardous waste	t
Economic	ECON1	Net revenue	R$MM
	ECON2	Adjusted EBITDA	R$MM
	ECON3	Total net margin	%
	ECON4	Net debt	R$MM
	ECON5	Gross kofit	R$MM
	ECON6	Gross margin	%
Social	SOC1	Number of employees	–
	SOC2	Lost time injury rate	%
	SOC3	Injury rate without leave	%
	SOC4	CRI deaths	–

Source Authors

1. Environmental: 302–1, 303–1, 305–1, 305–2, 305–3, 305–4, 306–2;
2. Economic: 103, 201–1, 203–1, 203–2;
3. Social: 102–8, 403–2.

The data regarding the dimensions and criteria for the two companies for the years 2018 and 2019 constituted data for the construction of the Evaluation Matrix. Table 7.4 presents the Evaluation Matrix for the study.

After the preliminary analyses, the normalization of the data for each criterion was performed, according to the two companies, i.e., the data used in the calculations, considered their homogeneity of parameters. The same weight is assigned for the different criteria, so that the sum of the respective scores had the unit value.

The ideal solution (PIS) and anti-ideal solution (NIS) were subsequently determined, and the Euclid distances and coefficients for each of the criteria were then determined, in relation to the two companies, both in the years 2018 and 2019. Table 7.5 presents the results.

The Technique of Ordering Preferences for Similarity with the Ideal Solution (TOPSIS) allowed the ordering of alternatives by coefficient, ranging from 0 to 1,

Table 7.4 Evaluation matrix

		ENT1		ENT 2	
		2018	2019	2018	2019
Environmental	ENV1	3,75,65,096.00	3,69,40,688.45	34,95,029.09	4,48,81,517.80
	ENV2	2,11,59,291.00	2,22,68,547.47	5,55,99,013.82	6,01,91,597.88
	ENV3	0	0	1,58,451.27	1,25,782.72
	ENV4	26,99,956.00	17,00,888.74	1,06,68,979.54	6,96,59,103.35
	ENV5	23,67,008.97	22,57,647.13	43,79,351.82	45,93,547.51
	ENV6	1,49,948.20	2,24,351.51	15,50,524.56	16,15,547.85
	ENV7	7,21,380.67	7,36,489.57	7,04,850.13	9,30,672.31
	ENV8	12,84,896.81	13,78,076.78	3,62,68,734.92	3,79,49,088.58
	ENV9	1,86,87,792.94	2,07,44,658.22	3,15,35,966.75	3,20,28,946.79
	ENV10	4,413.46	1,096.88	29,542.03	32.706.93
	ENV11	4,39,066.89	5,27,560.53	21,61,720.30	21,44,268.94
Economic	ECON1	30,188	33,447	1,81,680.20	2,04,523.60
	ECON2	2,462	5,317	14,849.80	19,881.10
	ECON3	−7%	3.60%	0.00%	3.00%
	ECON4	−15,610	−13,269	47,217.70	42,994.10
	ECONS	4,868	8,077	26,340.20	31,946.40
	ECON6	16.10%	24.10%	14.50%	15.60%
Social	SOC1	1,07,784	94,755	2,30,086	2,42,105
	SOC2	1.19	1.25	2.12	1.99
	SOC3	22	24.7	17.77	17.44
	SOC4	0	1	1	3

Source Authors

the best alternatives being those whose overall performance is closer to 1. Table 7.6 presents the final classification order of the criteria.

It is inferring that the highest global performances in relation to EMP1 are related to the "Economic" dimension. On the other hand, those of EMP2 are mostly related to the "Social" dimension.

EMP1 achieved higher performances in this dimension, as it was a publicly traded company, with shares traded on the stock exchange in Brazil (ISE B3– BRFS3) and in the United States (NYSE—ADR level III) and a market value of R$ 28.6 billion at the end of 2019 and has executed a restructuring plan aimed at recovering performance consistent with the potential of the business. On the other hand, its lower performances were related to the social dimension, contrasting with EMP2, which obtained its greatest performances in this dimension.

The low performances associated with the social dimension are related because in 2019, the adoption of the *occupational safety and health administration* (OSHA) criteria for classification of accidents, which also consider the classification of First

Table 7.5 Distance and coefficient measurements

		ENT 1			ENT2		
		2018	2019	c	2018	2019	c
ENV1	D+	0.41824	0.39517	0.24513	0.58914	0.99397	0.06002
	D−	0.04187	0.04391		0.08033	0.31969	
ENV2	D+	0.37148	0.44175	0.24508	0.79081	1.22502	0.0123
	D−	0.04594	0.03976		0.03264	0.1923	
ENV3	D+	0.88704	0.88704	0	0.54395	0.2313	0.05527
	D−	0	0		0.08144	0.10608	
ENV4	D+	0.68995	0.11672	0.23825	0.31767	0.97694	0.0129
	D−	0.01695	0.06196		0.0213	0.02523	
ENV5	D+	0.43916	0.37409	0.24509	0.35499	0.42138	0.05822
	D−	0.04	0.04572		0.09923	0.09381	
ENV6	D+	0.14716	0.65944	0.2402	0.35963	0.41677	0.05826
	D−	0.06096	0.01965		0.09888	0.09422	
ENV7	D+	0.39243	0.42097	0.24513	0.2017	0.57343	0.05389
	D−	0.04414	0.04163		0.1069	0.07809	
ENV8	D+	0.35844	0.45461	0.24502	0	0	1
	D−	0.04704	0.03859		0.0991	0.09396	
ENV9	D+	0.33479	0.47779	0.24488	0.37743	0.39902	0.05834
	D−	0.04899	0.03647		0.09751	0.09575	
ENV10	D+	0.94183	0.08448	1.5459	0.3176	0.45851	0.05775
	D−	0.00082	0.0335		0.10179	0.09035	
ENV11	D+	0.2812	0.52982	0.24429	0.39387	0.39387	0.05678
	D−	0.05312	0.03163		0.09618	0.0971	
ECON1	D+	0.37656	0.51136	0.48332	0.30394	0.46776	0.01209
	D−	0.36918	0.40776		0.04976	0.0356	
ECON2	D+	0.0334	0.82345	0.46074	0.19111	0.58011	0.01148
	D−	0.17655	0.44893		0.05775	0.02537	
ECON3	D+	0.79025	0.06491	0.46687	1	1	0
	D−	0.44787	0.20791		0	0	
ECON4	D+	0.54872	0.33661	0.48278	0.4509	0.32093	0.0031
	D−	0.4162	0.35512		0.00102	0.04202	
ECON5	D+	0.13186	0.7291	0.47429	0.25379	0.5175	0.0119
	D−	0.25703	0.44372		0.05357	0.03107	
ECON6	D+	0.18631	0.68187	0.47787	0.33516	0.43676	0.01215
	D−	0.28857	0.43867		0.04722	0.0384	
SOC1	D+	0.47481	0.29684	0.20823	0.38613	0.45533	0.34726
	D−	0.03496	0.01841		0.30251	0.31123	

(continued)

Table 7.5 (continued)

		ENT 1			ENT2		
		2018	2019	c	2018	2019	c
SOC2	D+	0.35178	0.42023	0.24993	0.46362	0.37767	0.34703
	D−	0.04583	0.03988		0.31208	0.30124	
SOC3	D+	0.30546	0.46625	0.24959	0.4336	0.4081	0.34763
	D−	0.04964	0.03574		0.30882	0.30561	
SOC4	D+	1	1	0	0.007	0.9	0.16354
	D−	0	0		0.1	0.30105	

Source Authors

Table 7.6 Order of classification of criteria

		ENT 1	ENT2
Environmental	ENV1	9°	6°
	ENV2	12°	14°
	ENV3	20°	12°
	ENV4	17°	20°
	ENV5	11°	9°
	ENV6	16°	8°
	ENV7	10°	13°
	ENV8	13°	1°
	ENV9	14°	7°
	ENV10	19°	10°
	ENV11	15°	11°
Economic	ECON1	1°	16°
	ECON2	6°	18°
	ECON3	5°	21°
	ECON4	2°	19°
	ECON5	4°	17°
	ECON6	3°	15°
Social	SOC1	18°	3°
	SOC2	7°	4°
	SOC3	8°	2°
	SOC4	20°	5°

Source Authors

Aid, with an increase in the rates of this category. Changes in indicators will be identified as controls are continued.

Meanwhile, EMP2 has achieved its greatest performance in relation to the social dimension, as its control is based only on regional programs and local action plans in the regions in which it operates, although it operates in 15 countries, and does not take into account the robustness of the criteria adopted by OSHA that EMP1 that operates in 130 countries adopts, thus affecting indirectly the results.

Although in 2019, EMP2 recorded the highest net revenue in the company's history, the reduction in indebtedness, the poor performance of the ECON3 criterion shows that the margin increase was much lower than expected and may be related to the transaction of approximately US$ 392 million in the acquisition of Tulip by Pilgrim's.

In general, after taking into account the two alternatives, i.e., the two companies for the TOPSIS Method, it was inferred that according to the coefficients obtained, the performances in 2019 were higher than that of 2018, both in EMP1 and INP2, and the overall performance of EMP2 (0.71812) was higher than that of EMP1 (0.62203), when using the *Triple Bottom Line model.*

Bringing as parallel the study of Campos et.al (2020), in this study were defined the environmental criteria (8), and socioeconomic criteria (3), as well as their impacts and weights, the alternatives (8) are defined according to interventional techniques. After the impact levels applied for the evaluation of each criterion are defined, the diagnosis of a watercourse condition according to the evaluation themes, both for Reach 1 [0–770 m] and Reach 2 [770–1740 m], the decision matrix with the impact levels according to the criteria, PIS and NIS, positive and negative distances, coefficient and final classification for the alternatives related to Reach 1 and Reach 2.

The Caiado et al. (2019) study presents the concept and application of the methodology of the Technique of Ordering Preferences by Similarity with the Ideal Solution (TOPSIS), presenting the stages of definition of objectives, matrix, weights and impacts, standardization, data weighting, determination of the ideal solution (PIS) and anti-ideal (NIS), calculations of Euclid distances (D + and D−), calculation of the coefficients for each alternative, enabling the ordering of alternatives by coefficient both in 2013 and in 2014.

In the study of Paz et al. (2021), after the steps for applying the method, the data are arranged in a decreasing way in the approximation coefficient and made available in the dashboard format with 3 screens that can display, for example, indicators by municipality, comparison between municipalities by sustainability dimension. This form of visualization makes it easier for decision makers to make decisions about social policies, such as health, garbage collection, illiteracy, among others.

The arrangement of the results in the form of dashboards facilitates consultation of the information, functioning as another tool to support decision makers as it filters the information and provides quick and attractive visualization of the results of the study.

7.7 Summary of Ideas and Main Concepts

- The data resulting from the research proved to be robust for the analysis of sustainability through the Triple Bottom Line in light of the Similarity Technique by Ideal Solution (TOPSIS) for managing animal protein companies. The companies studied are both Brazilian with plants in different countries and present in their respective reports the homogeneity in relation to the units and the GRI's studied.
- The method makes it possible to rank the global performance of the criteria, the companies and the years considered. Management implications in relation to results can be developed, aiming at benchmarking, in line with possible changes in some aspects of strategic planning and tactical planning of the organization.
- The existence of limitations of the concepts presented is notorious, mainly in regarding the different level of coverage and detail of the report for the two companies studied. If this level were more homogeneous, it would make it possible to obtain greater number of criteria to be analyzed.

It is suggested in future works, the increase in the number of alternatives, i.e., companies in the same sector, for a possible new approach to the food sector of animal protein, taking into account the rise in consumption of foods from of vegetable and plant-based protein.

7.8 Questions for Reflection and Study

(1) What is the relationship between the Sustainable Development Goals (SDGs) and the ESG methodology?
(2) What are the advantages of adopting practices related to the ESG methodology for the companies in the study in question?
(3) Summarize the strategies presented by the authors to analyze sustainable practices in Brazilian companies, in the animal protein sector, in more than one country?
(4) What were the main limitations found in the study and what can be done to improve them?

7.9 Cases to Be Resolved

(1) How can an organization add value to its processes (considering its value chain), to obtain good results in assessments (ratings) and remain well categorized in the benchmarking developed for sustainability criteria? How can the item "Governance" be fundamental in this task directly focused on the value chain?

(2) Which practical methods and which methods developed by the academy could be used by organizations to prioritize criteria to be analyzed? How could research centers collaborate with organizations in this regard? When the criteria are numerous, and the problems analyzed are complex, different methodologies can be confronted so that the most efficient is considered.
(3) How can an organization act to produce greater adherence between its sustainability report and the Sustainable Development Goals? Could this proactive action cooperate in improving institutional indicators? Does the institutional rating improve? How to promote adhesion between the tactical and strategic sectors to reflect on the operational?
(4) How to deal with the institutional challenge of connecting ESG with the Sustainable Development Goals knowing that companies have a relevant social role are increasingly required in relation to their positioning in the face of current dilemmas? Can these goals serve as support for planning changes and adopting strategic practices within the ESG field?
(5) Will the sustainable process to be adopted feedback not only from its own actions, but also by positively influencing its stakeholders?

References

ABIA - Associação Brasileira da Indústria de Alimentos (2021) Avaliable in https://www.abia.org.br/numeros-setor (Access in May 22 of 2021)
ABIEC - Associação Brasileira das Indústrias Exportadoras de Carnes (2021) Avaliable in http://abiec.com.br/exportacoes (Access in August 23 of 2021)
Ahmed M, Qureshi MN, Mallick J, Ben Kahla N (2019) Selection of sustainable supplementary concrete materials using OSM-AHP-TOPSIS approach. Adv Mater Sci Eng 2019(1):2850480
Almeida ABFR, Araújo MM (2013) The right to sustainable development and the symbolic dimension of its application. Belo Horizonte: Dom Helder Câmara Superior School
Amado FADT (2016) Outlined environmental law, 7th edn. Method, São Paulo
Barbieri JC (2016) Corporate environmental management: concepts, models and instruments, 4th edn. Saraiva, São Paulo
Bressanelli G, Perona M, Saccani N (2019) Challenges in supply chain redesign for the circular economy: a literature review and a multiple case study. Int J Prod Res 57(23):7395–7422
Brown HS, De Jong M, Levy DL (2009) Building institutions based on information disclosure: lessons from GRI's sustainability reporting. J Clean Prod 17:571–580
Caiado RGG, Quelhas OLG, Nascimento DLDM, Anholon R, Leal Filho W (2019) Towards sustainability by aligning operational programmes and sustainable performance measures. Prod Plan Control 30(5–6):413–425
Caiado RGG, Scavarda LF, Azevedo BD, De Mattos Nascimento DL (2022) Challenges and benefits of sustainable industry 4.0 for operations and supply chain management—a framework headed toward the 2030 agenda. Sustainability 14:830–856
Campos PCO, Paz TSR, Lenz L, Qiu Y, Alves CN, Simoni APR, Amorim JCC, Lima GBA, Rangel MP, Paz I (2020) Multi-criteria decision method for sustainable watercourse management in urban areas. Sustainability 12:6493
Carvalho NLA, Ribeiro PCC, Oliveira LK, Silva JEAR, Vieira JGV (2019) Criteria to implement UDCs in historical cities: a Brazilian case study. European Transport 72:1–29

Cattani AD, Laville JL, Gaiger LI, Hespanha P (Orgs.) (2009) Dicionário Internacional da Outra Economia. São Paulo: Almedina Brasil

Cinelli M, Coles SR, Kirwan K (2014) Analysis of the potentials of multi-criteria decision analysis methods to conduct sustainability assessment. Ecol Ind 46:138–148

Costa BS (2013) Environment as a right to life: Brazil, Portugal and Spain, 2nd edn. Lumen Juris, Rio de Janeiro

Costa FV, Lima GBA (2020) Survey of performance indicators. XL National Meeting of Production Engineering - ENEGEP 2020. [S.l.]: [s.n.]

Darnall N (2003) Why firms adopt ISO 14.001: an institutional resource-based review. In: Academy of management conference (2003, Seattle). Proceedings Briarcliff Manor: AOM

Donaire D (1999) Environmental management in the company, 2nd edn. Atlas, São Paulo

Elliott J (2013) An introduction to sustainable development (4th ed.) [S.l.]: Routledge

Elkington J (2011) Sustainability, knives and fork cannibals. Macron Books, São Paulo

Froehlich C (2014) Sustainability: dimensions and methods of measuring results. Unilasalle Manag Mag 3(2):151–168

Freitas J (2016) Sustainability: right to the future, 3rd edn. Forum, Belo Horizonte

GIRAC (1987) Strategie forecasting studies, BP-30. Switzerland, Geneva

Gomes MF, Ferreira LJ (2018) Public policies and sustainable development goals. Law Dev 9(2):155–178

Goyanes M, Demeter M, de Grado L (2020) The culture of free: construct explication and democratic ramifications for readers' willingness to pay for public affairs news. Journalism 23(1):207–223. https://doi.org/10.1177/1464884920913436 (Original work published 2022)

GRI–Global reporting initiative (2021) About GRI. Site GRI. Available in: https://www.globalreporting.org/about-gri/. Accessed in May 2021

Hwang CL, Yoon K (1981) Multiple attribute decision making. Economics and Mathematical Systems, Berlin

Marzouk M, Sabbah M (2021) AHP-TOPSIS social sustainability approach for selecting supplier in construction supply chain. Clean Environ Syst 2. Available in. https://www.sciencedirect.com/science/article/pii/S266678942100026X

Martins VWB, Anholon R, Quelhas OLG, Leal Filho W (2022) Roadmap to enhance the insertion of social sustainability in logistics systems. Int J Prod Perform Manag 1:1–21

Mafra JR (2015) The paradigm of sustainability in the Brazilian legal system: a fundamental material right. Electron J Law Polit 10(1):547–566

Milaré É (2014) Environmental law, 9th edn. Courts Magazine, São Paulo

Mueller G (1996) Leading companies, economic power and small producers in the Agroindustrial meat chain in Brazil. Econ Information—Government State São Paulo Inst Agric Econ 26:9–26

Norman W, MacDonald C (2004) Getting to the bottom of "triple bottom line." Bus Ethics Q 14(2):243–262

Ozkaya G, Erdin C (2020) Evaluation of smart and sustainable cities through a hybrid MCDM approach based on ANP and TOPSIS technique. Heliyon. Available in: https://www.sciencedirect.com/science/article/pii/S2405844020318958

Page MJ, Moher D, Bossuyt PM, Boutron I, Hoffmann TC, Mulrow CD, McKenzie JE (2021) The PRISMA 2020 statement: an updated guideline for reporting systematic reviews. Int J Surg 88. Available in: https://www.sciencedirect.com/science/article/pii/S1743919121000406

Pascoal LL et al (2011) Commercial relations between producer, industry and retail and the implications for differentiation and pricing of meat and non-carcass beef products. Braz J Anim Sci 40 (Suppl. Special)

Paz TSR, Caiado RGG, Quelhas OLG, Gavião LO, Lima GBA (2021) Assessment of sustainable development through a multi-criteria approach: application in Brazilian municipalities. REAd|Porto Alegre 27(1):128–152

Ribeiro BC (2020) Barriers in the implementation of a quality management system: action research in an animal protein industry. Completion of course work (Graduate in Production Engineering)—Federal University of Uberlândia, Ituiutaba

Révillion JPP, Kapp C, Badejo MS, da Veiga Dias V (2020) O mercado de alimentos vegetarianos e veganos: características e perspectivas. Cadernos de Ciência & Tecnologia 37(1):26603

Soltani MAH, Mahmoudi MD (2020) An integrated framework using VSM, AHP and TOPSIS for simplifying the sustainability improvement process in a complex manufacturing process. J Eng 18:211–229

Siqueira MR de (2016) The determinants of the internalization of Brazilian companies in the beef sector. Master's Thesis in Economic Sciences. Piracicaba

Torres LF, Anholon R, Rampasso IS, Quelhas OLG, Leal Filho W, Silva D (2022) Perception of shop floor employees regarding senior management support in lean projects and its relationship with initiatives success. Revista De Administração Da UFSM 15:1–14

Vogt M et al (2017) Relationship between determinant factors of disclosure of information on environmental impacts of Brazilian companies. Manag Stud 33(142):24–38. Available in: https://www.sciencedirect.com/science/article/pii/S0123592316300729

WCED–World commission on environment and development (1987) Our Common Future. Oxford University

Juliana Damaris Candido De Lima Doctoral Student in Production Engineering in the Management, Business Strategy and Finance concentration area at the Fluminense Federal University (UFF/Niterói), master's in production engineering in the Production Management concentration from the State University of North Fluminense Darcy Ribeiro (UENF) and Graduation in Production Engineering from the State University of North Fluminense Darcy Ribeiro (UENF). He was part of the Multicriteria Decision Aid Research Project and Multivariate Data Analysis applied to Service Quality Assessment (LEPROD/CCT/UENF). Acted as Substitute Professor in Service Management, in the Federal University of Rio de Janeiro (UFRJ/RJ) and also as Substitute Professor in Administrative Sciences Departments at the Rural University of rio de Janeiro (UFRRJ/RJ) and works as Assistent professor at the Department Production Engineering at the Federal University of Rio de Janeiro (UFRJ/RJ).

Renan Carriço Payer Doctoral Student in Production Engineering in the Management, Business Strategy and Finance concentration area at Fluminense Federal University (UFF/Brazil), master's in production engineering in the Management, Business Strategy and Finance concentration area at Fluminense Federal University (UFF/Brazil), Graduation in Chemical Engineering from the University of the State of Rio de Janeiro (UERJ/Brazil) and Graduation in Industrial Chemical at Fluminense Federal University (UFF/Brazil). Develops research in the areas of sustainability and circular economy. He is currently a federal public employee in a Brazilian state-owned company.

Samuel Martins Drei Ph.D. student in Production Engineering, as a CAPES scholarship holder, at the Fluminense Federal University, with a concentration in Systems, Decision Support and Logistics. Master's in production and manufacturing engineering, with a CAPES scholarship, from the State University of Campinas, at the Faculty of Applied Sciences, with an area of concentration in Operations Research and Process Management. Graduated in Production Engineering from the Federal University of Ouro Preto, at the Institute of Exact and Applied Sciences.

Lizzie Bessa Risicato Master's student in Production Engineering at the Fluminense Federal University (UFF/Brazil). Graduated in Production Engineering at the Federal Center of Technological Engineering Celso Suckow da Fonseca–CEFET-RJ (2008) and postgraduate in Project Management (2020). Worked in Engineering and Management Consulting companies, such as: GPI/UFRJ e Engineering and Management Institute. Expertise in research in Scientific Initiation projects. Author of the articles Multicriteria methodology in cost assessment in occupational safety, published in the journal Sistemas e Gestão of the Postgraduate Program in Management Systems at LATEC/TEP/TCE/PROPP/UFF.

Ruben Huamanchumo Gutierrez Titular Professor and Researcher of the Federal Fluminense Univ.Has a D,Sc. (1997) and MSc. (1990) degrees by COPPE/UFRJ (Brazil). He graduated in Industrial Engineering in 1985. Has business experience and teaching (30 years) in Quality Management, Process Improvement; Projects; Business Strategies; and financial issues; applying Multicriteria Decisions (MCDA), Systems Information Management, and scenarios. Published several books. Also published articles in relevant journals. Published and participated in events like ENEGEP ENANPAD, EPIO (Argentina), CORS (Canada) etc., also worked in scientific societies such as ABEPRO, SOBRAPO, CLADEA etc.

Lidia Angulo-Meza She has a degree in Operational Research Investigación Operativa—Universidad Nacional Mayor de San Marcos, a master's degree in production engineering from COPPE, Federal University of Rio de Janeiro and a Ph.D. in Production Engineering from COPPE Federal University of Rio de Janeiro. She is currently a full professor at Fluminense Federal University, in the Production Engineering Department, at the Fluminense Federal University's School of Engineering and is a CNPq Research Productivity Scholar.

Gilson Brito Alves Lima Graduated in Civil Engineering with specialization in Occupational Safety Engineering. Ph.D. in Production Engineering. Full Professor, Department of Production Engineering, Fluminense Federal University (UFF). Works in the areas of Teaching, Research, and Extension, through training projects and technological development, with emphasis on the following topics of Industrial Management: Safety, Environment, Risk, Maintenance.

Lincoln Campelo Dias Ph.D. student in Production Engineering at the Federal University of Rio de Janeiro, master's in engineering systems management at the Catholic University of Petrópolis, Brazil. Mechanical engineer and bachelor's in mathematics.

Pâmela De Carvalho Marques Silva Degree in Actuarial Sciences from the Federal University of Rio de Janeiro (2019) and a degree in Statistics from the Federal University of Rio de Janeiro (2019). Has experience in Probability and Statistics, with an emphasis on Statistics.

Chapter 8
ES(?)G—The Social Dimension as a Necessary Counterpoint to the Market's Desires

Fernanda Ribeiro Ramos Lopes, Fernando Oliveira De Araujo, Felipe Rodrigues Do Nascimento, and Wagner Dos Santos Ramos

Abstract This chapter delves into the social dimension of ESG criteria, critiquing its current limitations in prioritizing internal human resources and shareholder concerns over broader societal issues, particularly evident in developing countries like Brazil. It traces the historical evolution of Corporate Social Responsibility (CSR) from philanthropy to a more holistic management approach encompassing ethical, social, and environmental considerations. Emphasizing the importance of social responsibility in ESG evaluations, it highlights key factors impacting company operations, including job creation, diversity, community relations, and human rights. The chapter underscores the acknowledgment of social responsibility by financial institutions and investors. Empirical evidence, such as surveys, demonstrates the significance of the social aspect of ESG, particularly in the wake of events like the COVID-19 pandemic. As a contribution, it offers a comprehensive overview of the challenges and opportunities associated with integrating social concerns into business practices within the broader context of CSR and sustainability, addressing the evolving dynamics of the market and global sustainability agendas.

F. R. R. Lopes
Organizational Management, Management Systems Program, School of Engineering, Fluminense Federal University, Niterói, Rio de Janeiro, Brazil
e-mail: fernanda_ramos@id.uff.br

F. O. De Araujo (✉)
Organizational Management, Production Engineering Department, School of Engineering, Fluminense Federal University, Niterói, Rio de Janeiro, Brazil
e-mail: fernandoaraujo@id.uff.br; faraujo@sustentabile.com.br

F. R. Do Nascimento
Business Economics, Institute of Educational Development, Getulio Vargas Foundation, Rio de Janeiro, Brazil
e-mail: frnascimento@id.uff.br

W. Dos Santos Ramos
Fluminense Federal University, Niterói, Rio de Janeiro, Brazil
e-mail: wagnerramos@id.uff.br

M. J. Meiriño and O. L. G. Quelhas (eds.), *Critical Factors Associated with Environmental, Social and Governance Management in Organizations*, Accounting, Finance, Sustainability, Governance & Fraud: Theory and Application, https://doi.org/10.1007/978-981-96-1725-8_8

Keywords Corporate social responsibility · Corporate sustainability · ESG's social dimension · Creation of shared value · Social dimension of sustainability

8.1 Introduction

Welcome to a journey through the dynamic realm of ESG—Environmental, Social, and Governance—criteria. In this chapter, we shine a spotlight on the often-overlooked yet pivotal aspect of ESG: Its social dimension. As the global landscape of corporate responsibility evolves, we delve into the historical roots of corporate social responsibility (CSR) and its integration into the fabric of organizational governance.

Our exploration extends to the contemporary discourse surrounding ESG, where we dissect its fundamental components and implications for corporate conduct. Through a lens focused on the Brazilian context, we unravel the successes and challenges of ESG implementation, shedding light on its transformative potential and the imperative for a more nuanced understanding of social responsibility in the pursuit of sustainable development.

8.2 Role of this Chapter in the Book's Structure

The present chapter focuses on the relevant subject of ESG, particularly regarding this agenda's social approach. Within this context, there is a special focus on the fallacies and/or opportunities for perfecting the social dimension as considered by the ESG parameters, which in some cases limit the social agenda and internal human resource matters, frequently banalizing community issues—with a narrow focus, especially regarding poor communities located around the places of operation—and cultural matters—which are also relevant for a broader discussion about sustainability.

In this respect, this chapter presents the history of the social approach and discusses relevant concepts to the subject, besides presenting ESG social dimension success stories from Brazilian organizations recognized by the Corporate Sustainability Index (ISE) maintained by the [B^3], Brazil's Official Stock Exchange, which is headquartered in the city of São Paulo. The cases highlight the social aspects that, in practice, present themselves as indicators pertaining to internal human resources or to priority deliveries for shareholders, without any projection to neighboring communities and social projects.

Although ESG criteria are a power in terms of market articulation in favor of sustainability demands, it is possible to observe an operationalization of sustainability issues in their social dimension that is restricted to indicators with interests in social facts that are shown to be scarce considering the encompassing and interrelated view of sustainability.

Naturally, there is no expectation that the authors' view should supersede the criticism from readers regarding the applicability of the ESG instrument and the permeability of this practice for generating a change of posture that is strictly market-related for a more sensitive and solidary viewpoint regarding multiple social demands, especially in developing countries such as Brazil.

8.3 A History of the Subject's Development

Discussions on company responsibility became more intense in the second half of the twentieth century, challenging the traditional view based on maximizing shareholder profits and creating jobs, with the only limits for action being the practices and legal obligations of a balanced free market without fraud. The pressure to adopt socially responsible behaviors gained strength in light of the growing impacts generated by business activities. Initially, topics such as the environment, work relations, and consumer rights were among the most relevant along with philanthropic actions. The notion of corporate social responsibility (CSR), however, expanded to include broader social and public demands, such as communicating with and investing in neighboring communities, diversity and inclusion, and health and safety (Reis and Medeiros 2007).

Ashley (2005) states that CSR consolidated itself as the new ethos for business dealings and, in an allusion to Simon Zadek's viewpoint, highlights the importance of adopting ethical attitudes with publics of interest, such as respecting human rights and democracy, the environment, and committing to human and territorial development.

The most current definitions of the concept of CSR evolved from a piecemeal approach that is philanthropic in nature and external to the business to a perspective of integrating social and environmental agendas in a management system guided by the pillars of ethics and of relationships with interested parties (Gomes and Moretti 2007).

In this sense, the Ethos Institute, one of the most relevant organizations pertaining to responsible action in Brazil, understands CSR as "the form of management that is defined by an ethical and transparent relationship between the company and all the publics with which it interacts, and by the establishment of business goals that are compatible with the sustainable development of society, preserving environmental and cultural resources for future generations, respecting diversity, and reducing social inequality." To the Institute, it is vital to differentiate social responsibility from philanthropy, which is limited to piecemeal beneficent actions outside the company. CSR, on the other hand, is a management model that is applicable to all aspects of business, whose guidelines include having a relationship with all publics of interest (Instituto Ethos 2022).

In the international scenario, ISO Standard 26000 for social responsibility, which was launched in 2010, defines the concept as "an organization's responsibility for the impacts caused by its decisions and activities for society and the environment through ethical and transparent behavior that contributes to sustainable development, including the health and well-being of society; takes into consideration the interested parties' expectations; complies with applicable legislation; is consistent with international behavior standards; and is integrated throughout the organization and practiced in its relationships" (ISO 2010).

It is possible to infer that ethics and relationships with publics of interest are fundamental elements in CSR definitions. This has to do with managing the company's impacts on its interaction agents through the adoption of values and practices that support perfecting business processes from the ethical, social, and environmental standpoints (Tachizawa 2008).

Another pillar of CSR is the perspective of systemic management integrated with environmental and economic aspects. In this sense, the concept of sustainability becomes inseparable from responsible business management, as suggested by Elkington (1997) in the Triple Bottom Line, which assumes a balance between the social, environmental, and economic pillars in strategic business management. Elkington's ideas became the foundation for most of the discourse and strategies concerning corporate sustainability (CS).

CS assumes that there is not only a balance between social, environmental and economic aspects, but also the perspective of value generation, according to which a sustainable management must naturally generate a flow of benefits for all publics of interest (Savitz 2007). Currently, CS has been discussed according to the ESG agenda parameters. ESG stands for environmental, social, and governance, assembling sustainability criteria guided mainly toward investors and creditors in the evaluation and monitoring of impacts and risks.

The social dimension of sustainability or the ESG agenda is built from the same conceptual framework as CSR and assumes that companies establish management and social impact strategies in fields such as human rights, health and safety, labor, diversity and inclusion, dialogue, and social investment, among other elements related to human development and social well-being (Amato Neto 2011; Barbieri and Cajazeira 2016).

The evolution of companies' social actions is also reflected in the appearance of concepts such as the Social License to Operate (SLO), which includes strategies of dialogue and negotiation with impacted groups with the goal of obtaining the community's approval regarding the company's presence in their territory (Thomson and Boutilier 2011; Santiago 2016) and the creation of shared value (CSV), which proposes the redefinition of capitalism in a management proposal that correlates a company's competitivity with the well-being of the communities that surround it (Porter and Kramer, 2011). There is also the strengthening of businesses of impact, economically viable business initiatives that are created with the purpose of generating positive impact and offering solutions to resolve or reduce social and environmental problems (Yunus 2008).

The global debates and mobilizations regarding sustainable development, which currently include the Sustainable Development Goals (SDG) Agenda and the Principles from the UN Global Compact, also influenced the conceptual architecture of CSR and CS, and are important references for the incorporation of human and environmental values in business management strategies.

The conceptual structure of company responsibility is constantly evolving and has undergone transformations to respond to the complexity and intensification of sustainability challenges. In a recent motion, John Elkington proposed recalling the Triple Bottom Line. From his standpoint, after more than two decades, the concept was not capable of fostering the expected paradigm shifts. Originally, the term was conceived as the genetic code for changing capitalism into a logic of value generation for society and the biosphere. However, the author believes that a mistake was made in not recognizing the urgency and the effective potential of sustainability for changing reality, and he states that a new wave of innovation is needed, guided not only toward generating value, but also toward regenerating natural and human systems (Elkington 2018).

8.4 Basic Conceptual Structure—ESG and the Social Aspect

ESG connects the financial market and sustainability by incorporating environmental, social, and governance parameters in the financial evaluation of organizations. The term was first described in the 2004 report titled "Who Cares Wins," published by the United Nations Global Compact and the Swiss Federal Department of Foreign Affairs, in which the Secretary General of the UN at the time, Kofi Annan, provoked 50 CEOs from large financial institutions into reflecting about how to include environmental, social, and governance factors in the capital market (Pacto Global 2022).

ESG's social aspect stems from discussions originating in both Corporate Social Responsibility (CSR) and Corporate Sustainability (CS) regarding the aspects that are most strongly linked to matters of a social order. In the past, there was the idea that sustainability was only related to the environment, to ecology, and CSR had to do with social issues, such as human rights. However, CSR and CS are now considered synonyms (Van Marrewijk 2003).

Thus, the term Corporate Social Responsibility (CSR) is tied to companies' responsibilities to society and the environment and originates from the ideas that businesses and society are truly connected (Wood 1991) and that companies should promote social well-being beyond that which is required by law (McWilliams and Siegel 2001).

According to Font et al. (2016), this responsibility the companies have is noted by the economic, legal, and ethical characteristics of their activities compared to society's expectations regarding organizations. Furthermore, social and environmental responsibility happen through processes and tools that aim to create a Corporate Sustainability (CS) that involves the company's entire structure, including behaviors related to continuous improvement, diversity, differences, innovations, and ethics (Torelli 2021).

Ning et al. (2021) state that the financial industry has already formed a consensus on the ESG index and, thus, the analysis of environmental, social, and governance factors materializes a company's performance, risk, lucrativeness, and viability. According to the International Finance Corporation—IFC (2021), ESG are dimensions that companies consider when managing their operations, and also criteria analyzed by investors when they invest, pertaining to the business's risks, impacts, and opportunities. For the IFC (2021), social matters concentrate on but are not limited to potential or real changes for the neighboring community and works, such as health and safety, supply chain, and diversity and inclusion.

XP Investimentos (2022) considers that social factors are related to work relations and policies, inclusion and diversity, employee engagement, workforce training, human rights, community relations, privacy, and data protection. On their website, Marta Pinheiro, their ESG director informs readers that "the 'S' has stepped into the spotlight, especially considering the subject of diversity and inclusion, reaching the point of impacting governance." Table 8.1 summarizes social issues indicated by the IFC.

According to a study conducted by São Paulo's State Federation of Goods Trade, Services and Tourism (Fecomércio-SP, 2022), 46% of small and medium retailers consider the ESG's social aspect to be the most relevant point for their businesses. The Federation believes this perception could be related to the effects of the coronavirus pandemic, since many of the retailers made an effort to perform actions with positive social impact, such as not firing employees and adopting measures to control coronavirus dissemination among employees and customers.

Table 8.1 Summary of IFC social issues

Social Dimension
Company operations are conducted considering people's talents and abilities. Thus, the products and services companies offer through their operational activities involving production can either benefit or harm society
Creation of jobs and work conditions
Equal opportunities
Diversity
Training
Impacts on local communities
Health and safety
Child labor and forced labor in all supply chains
Complaint mechanisms
Human rights
Social impact of the company's products and services or operations
Gender violence and harassment

Source Created by the authors, adapted from IFC (2021)

According to research conducted by the National Industry Confederation (Messenberg 2021), the examples of ESG criteria from the agenda's social aspect that have been adopted by Brazilian industrial organizations based on what is relevant to the companies are: Health and Safety; Work Relations; Human Rights; Inclusion and Diversity; Community relationships; and Consumer relationships.

8.5 Cases of Relevant Companies in the ESG's Social Aspect

The São Paulo Stock Exchange has been conducting the Corporate Sustainability Index (ISE) ranking since 2005 (B3, 2022a). The methodology behind its criteria is based on an annual process, which involves a sectorized questionnaire, sending evidence, and other inclusion criteria (B3, 2022b).

The process includes the following dimensions: (i) Social Capital; (ii) Human Capital; (iii) Corporate Governance and High Management; (iv) Environment and Business Model and Innovation. Furthermore, the companies are considered according to their sectorial materiality, which is general (for all companies) and specific (varying according to the sector). The ISE score uses the Base-Score weighted by its Quality Factors. Within the ESG's Social aspect, the Human Capital and Social Capital guidelines are the base for everything (B3, 2022c).

In the Human Capital guidelines, there are general criteria: (a) Labor practices; (b) employee health and safety; and c) employee engagement, diversity and inclusion. As for the Social Capital guidelines, besides the general criteria, there are: (a) Human rights and community relationships; (b) private social investment and corporate citizenship; and (h) data security; as well as the specific criteria: (c) Technical and economic accessibility; (d) product quality and safety; (e) sales practices and product labeling; (f) customer well-being; and (g) customer privacy (B3, 2022d). Among the 73 companies that are listed, three of the five best-ranked companies stand out: EDP Energias do Brasil, Telefônica Brasil, and CPFL Energia.

EDP Energias do Brasil Case (EDP, 2019, 2020, 2021)

EDP Energias do Brasil has delivered various projects pertaining to diversity, employee inclusion, collective citizenship, and community relations. The company launched an internship program that reserves a quota of 50% for Black people and, to ensure the program is inclusive, they observe Black candidates' English levels and help improve it with a scholarship directed toward achieving proficiency in the language without requiring previous knowledge. This strategy increases the possibility of a diverse hiring process since, according to the Internship Company, 66% of Black and Brown students have no previous knowledge of the language. In 2021, the program benefited 22

Black interns and, for 2022, it was extended to all interns who are a part of the socioeconomically vulnerable profile contemplated by the initiative in 2021.

Another plan that has the capacity to increase diversity is the Electrician School, which is for women only. Besides supporting gender diversity, this initiative promotes community relations because the program is free. It also increases social investment and corporate citizenship because it provides a scholarship for the program's duration, learning materials, some of the students have the chance of getting hired at the end of the course. This initiative has already qualified more than 30 women to be professionals.

Another source of support: The guidelines are followed by the company through its EDP Institute, which sponsors 25 social projects from 24 different organizations. Through its "Solidary EDP" selection program, the company has invested 7.1 million Brazilian reais (BRL) for incentives in areas such as culture, childhood, teenage years, third age, and sports.

Still regarding diversity, the company altered its selection processes to give priority to candidates from underrepresented groups such as women, Black people, people with disabilities, trans people, professionals over 50 years old, and socially vulnerable people.

Since 2020, EDP began to include racial literacy education, specific training on subjects like gender, identity, and affective-sexual orientation. The training is directed to the company's leadership, besides the multiplier agents from the EDP Program for Diversity and Inclusion. The project is also in agreement with the corporate citizenship guideline since it discusses the impacts of structural racism on society and in organizations.

As for the diversity and social investment guidelines, EDP sponsors the Diversa Fair, the largest recruitment event for LGBTQI + youth in Latin America. It has also become a signatory of the Forum of Generations and Work Future commitments and promotes an annual Diversity Week, which is an internal event that stimulates reflections about diversity and inclusion.

Regarding the labor practices guideline, EDP joined the Empresa Cidadã Program, guaranteeing an extension of maternity leave from 120 to 180 days and paternity leave from 5 to 30 days (10 days past what the Program recommends). These benefits are also extended to homoaffective couples, and in case of adoption.

Despite the initiatives supporting diversity guidelines and being a part of the top five companies in the ISE ranking, the numbers show that there is still a lot to do. For instance, between 2019 and 2021, there was very little variation in the proportion of female collaborators: From 22 to 23%, while in leadership roles (management and administrative council) the variation was more noticeable, going from 12 to 19%. As for the proportion of Black collaborators, the variation was also low, going from 22 to 24%, with management-level seeing a shift from 6 to 7%. As for employees above 50 years old, the variation was negative and the proportion decreased from 12 to 11%.

In the labor practices guideline, the numbers also show some results that generate concern, since work accidents went up by 41% at EDP, with 42 more cases occurring between 2019 and 2021. The number of deaths also increased, from one death to five.

As for the social investment and community relations guidelines, the variation was significant. External social investment went up by 71%, jumping from 14 million BRL to 23 million BRL. If only the company's capital is considered, the variation was of 439%. A point of interest lies in the number of beneficiaries. The number of direct and indirect beneficiaries went down, from 80,758 to 66,334 and from 241,274 to 185,817, respectively.

Telefônica Brasil (Vivo) Case (Telefônica, 2019, 2020, 2021)

The company has ambitious goals, if compared with the structure of the current market. In the next triennium, Vivo intends to have at least 30% of Black professionals and 33% of female professionals in leadership roles.

Regarding the diversity guideline, some of the programs the company employs are: The Manifest for Diversity, Vivo Afro, Black People in Leadership, Raça Negra Prize, Women in Technical Fields, and the Specific Affinity Group. As for the Private Social Investment and Corporate Citizenship guidelines, the company has programs in place such as its educational itinerary for data science, various projects to improve public education through the Telefônica Vivo Foundation and it encourages the retention of professionals with physical disabilities.

The manifest for diversity is the company's commitment, made public as a part of the subject of diversity. The company is a part of the LGBT + Rights Forum. The goal is to combat discrimination against people who are gay, lesbian, bisexual, transvestites, trans, and intersex. This project is a part of the UN's Free and Equal program.

In 2021, the company reserved 50% of its work slots for Black students through an affirmative action internship. There is also the Vivo Afro project, which consists of development workshops based on ideas and suggestions for the evolution of racial representativity. Another program, Black People in Leadership, was designed to prepare Black professionals to occupy leadership roles. It lasts six months and helps build the professionals' image and self-knowledge to strengthen their hard and soft skills, and is conducted with the help of the DISC[1] leaders and evaluation academy. In 2021, 100 collaborators were contemplated with participation in this program.

The Raça Prize is promoted by the NGO Afrobras in a partnership with Zumbi dos Palmares University and Vivo. It is awarded to people who contribute to the fight against racism and in favor of social inclusion. In 2021, the prize was awarded by the CEO to an employee who was the victim of racism from a customer in one of Vivo's stores.

Another diversity project, this time involving gender, is called Women in Technical Fields. It has the goal of boosting women's presence in the Customer Service field, which is mostly occupied by men. This project is currently present in cities from 15 states. There is also the Specific Affinity Group, which is geared toward the company's LGBTI + collaborators and aims to help with the inclusion of certain agendas within the company.

In agreement with the Community Relations, Private Social Investment, and Corporate Citizenship guidelines, Vivo launched the "Black Screens" project. It consists of a digital exhibit of Black artists. The initiative brought the artists together for an exhibit containing original works. The program transformed 240 Vivo stores into galleries.

In numbers, it seems that the variations did not evolve in a linear manner along with the guidelines. Currently, there is increased attention paid to policies geared toward racial issues when compared to the previous triennium. The company's 2019 report made no mention of the number of Black employees and the word "Black" was mentioned exactly once in the entire report. In 2021, there is already knowledge that 30.28% of the company is composed of Black people and the word "Black" was mentioned fourteen times. Considering the variation in percentage shown in 2020, it is clear that the variation in the number of Black people between 2018 and 2021 was significant, increasing from 20.70 to 30.28%. However, the number of Black people in leadership roles did not experience the same magnitude of variation, increasing from 16 to 19%.

Regarding women and diversity, the numbers show a different story. There was a lower variation in the number of women employed by the company, from 42% (2019) to 43% (2021). However, there was a more significant variation of the percentage of women in leadership roles, from 22 to 29% and on the Administrative Council, from 25 to 33%. On the other hand, the variation in the number of LGBTQI + collaborators changed very little, from 3 to 4%.

As for the Human Capital in Labor Practices, Health and Safety guideline, the accident rate decreased from 0.33 to 0.26% and accident severity went down from 11.5 to 1.85%.

There are still some programs that cannot be evaluated through numbers due to their recent launch, such as the Technical and Professional Education Itinerary for Data Science, which aims to democratize access to data knowledge, giving access to all students from the public education network; School for All (Escola para Todos), which promotes anti-racism; the Wise Prize from the Profuturo Program and other prizes, such as the World Summit Awards.

When observing the numbers, the Vivo case presents a mixed report, despite not showing a variation of more than 5% in the number of Black people in leadership roles and total women employed, since there was a higher variation when compared to the total number of Black employees and women in leadership roles. There was also a higher variation in the numbers pertaining to work safety and pro-community initiatives, which can bring about relevant numbers. One point that deserves attention is the slight drop in the number of partnered suppliers in the triennium, from 1247 to 1230, and the decrease in the average amount of money invested in employee training, from 1400 to 1024 BRL, which pertains to guidelines that are also a part of the ESG's human and social capital policies.

CPFL Energia Case (CPFL, 2019, 2020, 2021)

The company manages the project through CPFL Institute. It is possible to observe that its current projects are more closely aligned with the ISE dimension of Social Capital, especially regarding the matters of community relations, private social investment, and corporate citizenship. The CPFL in the Hospital program is compatible with this subject and has been selected as a good practice of the UN's SDGs.

The program is divided into two aspects: Hospital Humanization and Hospital Infrastructure and Research Improvements. The first aspect, Hospital Humanization involves partnerships with initiatives such as: Médicos Sorrisos (Doctor Smiles), Leiturinhas nos Hospitais (Reading in Hospitals), and others that help create a humane hospital space to contribute to patients' recovery in hospital environments. The second aspect, Hospital Infrastructure and Research Improvements, includes initiatives from the Group's energetic efficiency sector that contribute to the structure of hospitals located in communities with ties to the Group. The work includes repairing existing installations, purchasing equipment, expanding spaces, and other activities related to improving the hospitals' infrastructure. Besides infrastructure support, CPFL also contributes to clinical research in the oncology field.

Another program related to the aforementioned guidelines is CPFL Jovem Geração (Young Generation). The Group seeks to reduce the levels of vulnerability in the communities to which they are connected by stimulating children and teenagers through sport and culture. In 2022, the company intends to work with institutions from 54 cities in states from three regions in Brazil: South, Southeast, and Northeast. Activities conducted by Young Generation include CPFL Young Generation at School, which works with children during the shift opposite to their regular schooling, thus allowing them to have full-time education. The program includes activities encouraging reading, sports, and cultural activities. In 2022, the plan is for the project to occur in 20 cities in the South and Southeast, impacting around 1900 lives.

Another example of a project related to the subject's guidelines is the Carreta Literária (Literary Truck). The program goes through different states with literary exhibits. In 2022, around 16 cities from five states were reached by the program with a collection of more than 500 books. This initiative complements the Semear (Planting) program which recruits CPFL Group volunteers to projects that benefit schools. The Semear program's objective is different from that of the Literary Truck, and involves planting community gardens, building computer labs, and providing art, dance, music, and theater lessons, among other actions alongside the community.

The company also has the Cinesolar program, which uses open air movie sessions using solar power to take culture to communities with which the company interacts. The project is taken to different regions in the format of a high technology traveling movie theater transported inside a solar-powered truck. The program has the capacity to seat 88 people per viewing.

In the same line, there are also two circuit projects: CPFL Circuit, which is a high adherence program bringing physical activities to the communities and the CPFL Energy Circuit, which conducts street races and walking circuits in different cities in the South and Southeastern regions of Brazil. The projects impact around two thousand people in each city. It is also important to mention the CPFL Philosophical Coffee: A TV show that is completely produced by CPFL and aired on TV Cultural since 2003. This show discusses topics such as diversity, inclusion, sustainability, and other cases related to social culture.

Despite the CPFL case being a part of the three cases of excellence according to the ISE ranking, a quantitative observation of the guidelines shows how the social aspect is still in its early days. Regarding diversity, the subject of diversity and inclusion only started to appear in the company's annual reports in 2020, which makes it difficult to make a triennial comparison such as those conducted with the previous companies.

Upon exploring some of the general axes, it is possible to note that the gender balance improved by 27.68%, going from 23.19 in 2019 to 29.61 in 2021. On the other hand, the numbers regarding gender identity and ethnic inclusion are not as clear.

As for the projects conducted by its institute, CPFL in Hospitals, for instance, had the goal of benefiting 200 hospitals and in 2021 it reached its goal, benefiting a total of 204. In 2022, the goal is to reach 327 hospitals. The CPFL Circuit impacted 79 thousand people in 2019 and 25 thousand in 2021. The pandemic may have played an important role in this reduction. Another number that deserves attention is that of the CPFL Institute's social investments: While direct investments remained stable with values of 3.7 and 3.8 million BRL invested in the period, the indirect investments experienced a relevant decline of -22.76%, going from 35.6 to 27.5 million BRL. As for the topic of Work Safety, the number of deaths resulting from work injuries and severe injuries was expressive, going from 0 to 2 and 41 to 49 cases, respectively.

8.6 Indication of How the Conceptual Structure Will Be Incorporated into the Development, Implementation, and Monitoring Process of the ESG's Social Aspect

Considering a scenario in which the financial paradigms that back the definition of ESG criteria can limit the reach and effectiveness of actions, especially in the social dimension, here is a list of suggested practices based on the previously mentioned literature with the potential to bring consistency to the social agenda in its broader and most fundamental goals on behalf of well-being and social justice.

- Establishing ethics-based relations with all publics of interest.
- Guaranteeing the high leadership's commitment to defining and conducting the ESG social dimension's strategies.
- Internally disseminating the social agenda and its concepts to encourage an organizational culture that is favorable to reaching and maintaining a satisfactory social performance with all impacted groups.
- Instituting a management system for Social Responsibility that includes parameters for planning, implementation, monitoring and improvement.
- Engaging in business initiatives, pacts and commitments connected to the relevant social issues.
- Promoting a constant dialogue with the publics of interest to learn about their expectations and to encourage engagement in the construction of relationship strategies with these groups.
- Defining the materiality of the social criteria inherent to the profile, the activities and the specific business context.
- Declaring public goals related to the social issues defined as priorities.
- Conducting a diagnostic of social action impacts and strategies.
- Defining realistic and measurable goals related to the social criteria defined as priorities in the materiality process.
- Selecting indicators related to the established goals for monitoring the evolution of the prioritized social actions.
- Designing management strategies and implementation processes that guarantee the effective integration of social criteria in business activities.
- Establishing multisector partnerships to strengthen the defined social actions.
- Reporting all positive and negative results of the company's social actions in a consistent, true, and accessible manner.
- Aligning the company's social goals with the UN's Sustainable Development Goals.

8.7 Summary of the Main Ideas and Concepts

- Corporate Social Responsibility—Has to do with the responsibilities of companies toward society and the environment, with the notion that business and society are intrinsically linked. Thus, companies should provide some sort of social well-being and make voluntary public commitments that go beyond those required by the law.
- Corporate Sustainability—Composed of the means that use processes and tools to perpetuate social responsibility; that is, the companies' responsibility includes the ethical, legal, and economic characteristics of their activities, besides society's expectations regarding what the company delivers to society. The company's entire structure is involved and employing behaviors related to continuous improvement, diversity, differences, innovation, and ethics.
- ESG's Social Dimension—Main social practices observed by the companies: Cultural incentives, social investment, customer relations, labor policies and relations, inclusion and diversity, employee engagement, workforce training, human rights, community relations, privacy, and data protection.
- Materiality—Process of defining the most important social, environmental, and governance issues considering the characteristics and context of each organization.
- Publics of Interest—All publics that affect or are affected by business activities.
- Transparency—Accountability practices that allow an organization to take responsibility and account for their activities for all publics of interest.
- Social Dialogue—Transparent and systematic communication strategies adopted by companies to establish conflict discussion and negotiation channels with legitimate representatives of the communities and publics affected by their activities.
- Private Social Investment—Voluntary investments made by the companies with their own resources in a planned, monitored, and systematic manner that is aligned with their business, considering their social goals of public interest.
- Philanthropy—The occasional, unplanned voluntary transference of resources, generally directed toward humanitarian aid and aligned with the business leadership's social affinities.
- Social Impact—The impacts of a social nature caused by the business activities in their publics of interest. Social impacts can happen directly, as is the case with surrounding communities, or indirectly, such as the impacts caused by suppliers and in society. It is vital for companies to establish trustworthy methods and systems to monitor, manage, and mitigate social impacts.
- Human Rights—These have to do with basic principles of respect and the promotion of rights and freedoms considered vital for human dignity.
- Diversity and Inclusion—A set of business strategies and practices that aim to create the same characteristics of diversity present in society in the company's group of collaborators, while also guaranteeing that all publics will have the same opportunities for development. The concept of diversity and inclusion also applies

to all social practices performed by a company regarding the external publics with which it has a relationship.

- Social License to Operate—Strategies for establishing a dialogue and negotiating with impacted groups to obtain the community's approval of the company's presence in the area.
- Creation of Shared Value—Management model that correlates a company's competitivity with the well-being of the surrounding communities and the fair distribution of resources among all groups with which the company interacts.

8.8 Questions for Reflection and Study

(1) Which are the positive impacts generated by the companies' adoption of social practices?

(2) Which are the negative impacts generated when a company invests in social practices with the sole purpose of obtaining tax incentives, with the clear intention of receiving benefits in exchange?

(3) Does making voluntary commitments in the social dimension contribute to a company's attraction and retention of talents with the aim to impact its productivity?

(4) Critique the main social practices of the companies cited in the cases and the benefits they afford society.

8.9 Cases to Solve

(1) Imagine an organization has more women in leadership roles and in its group of collaborators due to a coincidence and that it uses this indicator to advertise that it performs social practices that aim to include women in leadership roles. What can be done for this company to go from theoretical discourse to the effective practice of including women in leadership positions?

(2) Suppose that a company adopts business practices to comply with the racial quota law, but does the bare minimum required by the law, is not proactive, does not invest any resources in the behavioral, cognitive and relationship aspects pertaining to the interaction between collaborators who are persons with disabilities (PCD) and the other collaborators in the organizational context. With this, the PCDs end up feeling discriminated against at work, and this causes negative emotional consequences. What must this company do to revert the situation?

References

Amato Neto J (Org) (2011) Sustentabilidade & produção: teoria e prática para uma gestão sustentável. Atlas, São Paulo

Ashley PA (2005) Ética e responsabilidade social nos negócios, 2a edn. Editora Saraiva, São Paulo

B3—Brasil, Bolsa, Balcão (2022a) Índice de Sustentabilidade Empresarial (ISE B3). B3 (s.d.). Available at: https://www.b3.com.br/pt_br/market-data-e-indices/indices/indices-de-sustentabilidade/indice-de-sustentabilidade-empresarial-ise-b3.htm. Access 04 Jul 2022

B3—Brasil, Bolsa, Balcão (2022b) Metodologia do Índice de Sustentabilidade Empresarial (ISE B3) (s.d.). Available at: https://www.b3.com.br/data/files/DB/B2/66/3C/6B6AA71096B63AA7AC094EA8/ISEMetodologia-pt-br%20vf.pdf. Access on 04 Jul 2022

B3—Brasil, Bolsa, Balcão (2022c) ISE B3—Índice de Sustentabilidade Empresarial (Apresentação de Slides). B3 (s.d.). Available at: https://www.b3.com.br/data/files/36/65/97/55/50ED0810C493CD08AC094EA8/ISE%20B3%202022%20-%20Site.pdf. para citar formulação ISE. Access on 04 Jul 2022

B3 - Brasil, Bolsa, Balcão. (2022d). Diretrizes do Índice de Sustentabilidade Empresarial (ISE B3). B3. [s.d.] Available at: https://www.b3.com.br/data/files/F6/B2/FB/2C/6B6AA71096B63AA7AC094EA8/Diretrizes%20do%20ISE.pdf. para citar manual de diretrizes. Access on 04 Jul 2022

Barbieri JC, Cajazeira JER (2016) Responsabilidade social empresarial e empresa sustentável: da teoria à prática, 3ª. FGV, Rio de Janeiro

CPFL Energia (2019) Relatório anual 2019. CPFL Energia. 25/03/2020. Available at: https://cpfl.riweb.com.br/Download.aspx?Arquivo=LJ+KQbfONRZowl0UBTBGnQ==. Access on 14 Jul 2022

CPFL Energia (2020) Relatório anual 2020. CPFL Energia. 18/03/2021. Available at: https://cpfl.riweb.com.br/Download.aspx?Arquivo=x/FSwaDaFyb2Gy7NQ64kkw==. Access on 14 Jul 2022

CPFL Energia (2021) Relatório anual 2021. CPFL Energia. 28/03/2022. Available at: https://cpfl.riweb.com.br/Download.aspx?Arquivo=wjWODW6HHViMfy6ESb8XeQ==. Access on 14 Jul 2022

de Reis CN, Medeiros LE (2007) Responsabilidade social das empresas e balanço social: meios populares do desenvolvimento econômico e social. Editora Atlas, São Paulo

EDP Brasil (2019) Relatório anual 2019. EDP Brasil (s.d.). Available at: https://ri.edp.com.br/pt-br/informacoes-financeiras/relatorios-anuais/. Access on 28 Jun 2022

EDP Brasil (2020) Relatório anual 2020. EDP Brasil (s.d.). Available at: https://ri.edp.com.br/pt-br/informacoes-financeiras/relatorios-anuais/. Access on 28 Jun 2022

EDP Brasil (2021) Caderno ESG—Relatório anual 2021. EDP Brasil (s.d.). Available at: https://ri.edp.com.br/pt-br/informacoes-financeiras/relatorios-anuais/. Access on 28 Jun 2022

Elkington J (1997) Cannibals with forks—the triple bottom line of 21st century business. Capstone, Oxford

Elkington J (2018) 25 years ago I coined the phrase "Triple Bottom Line." Here's why it's time to rethink it. Harvard Business Review. Available at: https://hbr.org/2018/06/25-years-ago-i-coined-the-phrase-triple-bottom-line-heres-why-im-giving-up-on-it. Access on 25 Nov 2020

Fecomercio-SP—Federação do Comércio de Bens, Serviços e Turismo do Estado de São Paulo (2022) Pesquisa avalia aderência do comércio ao ESG. Fecomercio-SP. 21/02/2022. Available at: https://www.fecomercio.com.br/noticia/pesquisa-avalia-aderencia-do-comercio-ao-esg. Access on 02 Jul 2022

Font X, Guix M, Bonilla-Priego MJ (2016) Corporate social responsibility in cruising: using materiality analysis to create shared value. Tour Manage 53:175–186. Available at: https://doi.org/10.1016/j.tourman.2015.10.007. Access on 10 Sep 2021

Gomes A, Moretti S (2007) A responsabilidade e o social: uma discussão sobre o papel das empresas. Editora Saraiva, São Paulo

IFC—International Finance Corporation (2021) IFC ESG guidebook. IFC, Washington DC

Instituto Ethos (2022) Indicadores Ethos. Instituto Ethos (s.d.). Available at: www.ethos.org.br. Access on 26 Jun 2022.

ISO—International Organization for Standardization (2010) ISO 26000—Diretrizes em Responsabilidade Social. Inmetro website (s.d.). Available at: http://www.inmetro.gov.br/qualidade/responsabilidade_social/pontos-iso.asp. Access on 09 Aug 2022

McWilliams, A., & Siegel, D. (2001). Corporate social responsibility: A theory of the firm perspective. Academy of Management Review, 26(1), 117–127. Available at: https://doi.org/10.5465/amr.2001.4011987 Access on 25 aug. 2021.

Messenberg M (2021) Consulta ESG e a Indústria Brasileira. (Slideshow). Portal da Indústria (s.d.). Available at: https://static.portaldaindustria.com.br/media/filer_public/c9/e3/c9e3bc1f-b92a-4066-80cd-f5a5ed9c72de/ppt__consulta_esg_cni_3.pdf. Access on 28 Jun 2022

Ning X, Yim D, Khuntia J (2021) Online sustainability reporting and firm performance: lessons learned from text mining. Sustainability 13(3):1–15. Available at: https://doi.org/10.3390/su13031069. Access on 03 sep. 2021.

Pacto Global (2022) ESG. Pacto Global Brasil (s.d.). Available at: https://www.pactoglobal.org.br/pg/esg. Access on 26 Jun 2022

Porter M, Kramer MR (2011) Criação de valor compartilhado. Harv Bus Rev. Available at: http://hbrbr.com.br/criacao-de-valor-compartilhado/. Access on 15 Jul 2022

Santiago ALF (2016) Licença Social para Operar: relacionamento da empresa com a comunidade local. Tese (Doutorado em administração)—Centro Universitário FEI, São Paulo

Savitz AW (2007) A empresa sustentável: o verdadeiro sucesso é lucro com responsabilidade social e ambiental. Afonso Celso da Cunha Serra (trad.). Elsevier, Rio de Janeiro

Tachizawa T (2008) Gestão ambiental e responsabilidade social corporativa: estratégias de negócios focadas na realidade brasileira – 5 ed. revista e ampliada. Editora Atlas, São Paulo

Telefônica Vivo (2019) Relatório de Sustentabilidade 2019. Telefônica Vivo (s.d.). Available at: https://ri.telefonica.com.br/pt/esg/relatorios-de-sustentabilidade. Access on 04 Jul 2022

Telefônica Vivo (2020) Relatório de Sustentabilidade 2020. Telefônica Vivo (s.d.). Available at: https://ri.telefonica.com.br/pt/esg/relatorios-de-sustentabilidade. Access on 04 Jul 2022

Telefônica Vivo (2021) Relato Integrado 2021. Telefônica Vivo (s.d.). Available at: https://ri.telefonica.com.br/pt/esg/relatorios-de-sustentabilidade. Access on 04 Jul 2022

Thomson I, Boutilier RG (2011) Social license to operate. In: Darling P (ed) SME mining engineering handbook. Society for Mining, Metallurgy and Exploration, Littleton, pp 1779–1796

Torelli R (2021) Sustainability, responsibility and ethics: different concepts for a single path. Soc Responsib J 17(5):719–739. https://doi.org/10.1108/SRJ-03-2020-0081. Acesso em: 08 Sep 2021

Van Marrewijk M, Were M (2003) Multiple levels of corporate sustainability. J Bus Ethics 44:107–119

Wood DJ (1991) Corporate social performance revisited. Acad Manage Rev 16(4):691–718. Available at: https://doi.org/10.5465/amr.1991.4279616. Access on 04 Sep 2021

XP Investimentos (2022) ESG. XP Investimentos (s.d.). Available at: https://conteudos.xpi.com.br/esg/. Access on 02 Jul 2022

Yunus M (2008) Um mundo sem pobreza: a empresa social e o futuro do capitalismo. Editora Ática, São Paulo

Fernanda Ribeiro Ramos Lopes Has a master's degree in management systems from Universidade Federal Fluminense (2022) in the Social Responsibility and Sustainability Systems research line. She has a post-Graduate degree in Integrated Organizational Communication from the Escola Superior de Propaganda e Marketing (2017), an MBA in Business Management and Development with an emphasis on People Management from Universidade Federal do Rio de Janeiro (2010) and bachelor's degrees in journalism (2007) and Public Relations (2012) from Faculdades Integradas Hélio Alonso. She has been working in a professional capacity at the National Confederation of Goods Trade, Services and Tourism (CNC) in their Sustainability Program.

Fernando Oliveira De Araujo Has a post-doctoral degree in Production Engineering from UFRGS - Universidade Federal do Rio Grande do Sul. He has a PhD in Production Engineering from PUC-Rio - Pontifícia Universidade Católica do Rio de Janeiro and a master's degree in management systems and production engineering from UFF – Universidade Federal Fluminense. He is a professor, researcher, and extension program director in the Production Engineering department at Universidade Federal Fluminense and a lead researcher at NECSOS – Núcleo de Economia Criativa, Stakeholders, Organizações e Sociedade (DGP/CNPq) as well as at NEO - Núcleo de Empreendedorismo e Operações (DGP/ CNPq). He is also a registered researcher in the Management Systems Post-Graduate Program. He develops and participates in teaching, research, and extension initiatives, basing his interventions on a comparative and multidisciplinary methodological approach, conciliating production engineering knowledge with topics related to logistics and operations and services management, project management, change and stakeholder management, creative economy, innovation, and entrepreneurship.

Felipe Rodrigues Do Nascimento Has a master's degree in Economic Theory in the Macroeconomy research line from Universidade Federal Fluminense (2018). He has a bachelor's degree in economy from Universidade Federal Rural do Rio de Janeiro (2015) and is a professional investor (2012). He currently teaches at Fundação Getúlio Vargas and works as an economist for the BankRio Financial Group.

Wagner Dos Santos Ramos Has a PhD in Sustainable Management Systems with a line of research in Management of Sustainable Organizations (2021), a master's in management systems from Universidade Federal Fluminense (2013) with a line of research in Sustainability and Social Responsibility and an MBA in Organization and Strategy from the School of Engineering of the Fluminense Federal University (2011) with an emphasis on Social Responsibility. Graduated in Business Administration from Universidade Federal Fluminense (2009), he has been working professionally since 2009 in the areas of Corporate Social Responsibility and Sustainability.

Part V
Overlooking to ESG

Chapter 9
Environmental Consequences of CO_2 Emissions from the Use of Information Technologies in Brazil in the 2010–2021 Period

Priscilla Cristina Cabral Ribeiro

Abstract Most CO_2 emissions have their origin in the fossil fuels for energy and materials production, the other part, from the deforestation and wildfires. Brazil is one of the countries that are in the top list of emitting countries. The question is to conciliate the use of technologies and the CO_2 emissions decrease, considering their use that leads to an increasing of energy consume. One set of these technologies is the Information and Communication Technologies (ICT), which are the basis of process and management of all sectors in the world economy. Environmental, Social and Corporate Governance (ESG) allows companies to change their behaviours, planning, producing and selling in a more conscious way, according to a sustainable production. There is a debate where two theories relate the CO_2 emissions and the use of ICT. This chapter aims to fill the gap of this discussion, developing a discussion of ICT and sustainability through the debate over CO_2 emissions regarding these important technologies in the global economic context. The chapter will present a summary of literature and the suggestions to solve the problem, the classic conceptual framework with the main themes (ESG, ICT and CO_2 emissions), CO_2 and ICT Emission Cases, the practical use of the conceptual framework into the process of development, implementation and monitoring of the strategy and production organization, a summary of main ideas and concepts, questions and cases to be sold.

Keywords ESG · CO_2 emissions · Information and Communication Technology · Brazil

P. C. C. Ribeiro (✉)
Production Engineering Department at Universidade Federal Fluminense (UFF), Rio de Janeiro, Brazil
e-mail: priscillaribeiro@id.uff.br

M. J. Meiriño and O. L. G. Quelhas (eds.), *Critical Factors Associated with Environmental, Social and Governance Management in Organizations*, Accounting, Finance, Sustainability, Governance & Fraud: Theory and Application, https://doi.org/10.1007/978-981-96-1725-8_9

9.1 Role of the Chapter in the Structure of the Book

In 2015, UN member countries adopted the 2030 Agenda for Sustainable Development, which includes 17 Sustainable Development Goals (SDG). Built on the principle of 'leaving no one behind', this agenda is a universal call for all countries to act in favour of ending poverty, protecting the planet and increasing prosperity by 2030 (Alataş 2021). Closely related to this understanding, Information and Communication Technologies (ICT) has shown itself to be a promising path for Sustainable Development (Magazzino et al. 2021), despite the differences among the authors.

According to Dhakal (2011, p. 552), the term *governance* expresses the change from a traditional government structure towards a mechanism in which the State values actions of the private sector (market) and the third sector (civil society) in the process of decision-making. In recent years, the awareness and collective concern of Environmental, Social and Corporate Governance (ESG) has grown, resulting in a change of attitude throughout society, leading individuals to rethink their behaviour and impacts on the environment (Aich et al. 2021).

Starting from the discussion on themes related to Environmental, Social and Governance (ESG) in this book, this chapter will develop the discussion of Information and Communication Technologies and sustainability through the debate over CO_2 emissions regarding these important technologies in the global economic context.

9.2 History of the Development of the Theme

Two groups of authors were observed in the discussion of ICT and CO_2 emissions: those who concluded that the technologies would cause greater CO_2 emissions and others who understood that these technologies could reduce such emissions. In Table 9.1, a summary of the literature between the years 2010 and 2021 is presented, together with suggested solutions by the authors, which will be further discussed in chapter.

9.3 Classic Conceptual Framework—ESG, ICT and CO_2 Emissions

Sustainability is often presented in the literature as a reference to achieve economic development and growth with a balanced use of resources. Based on the dimensions (environmental, economic and social) proposed by Elkington (1998) and Sachs (1993), several authors have discussed, as presented in the previous items of this chapter, the consequences of using ICT.

Table 9.1 Summary of the literature on ICT and CO_2 emission (2010–2021)

Author	Country/Region	Model/Method	Does ICT emit CO_2?	Suggestions
Miyashita et al. (2010)	Kawasaki/Japan	–	Yes	Building management system Use of durable painting on the external parts of the building Water treatment Food and beverage pressing Teleconferences and videoconferences
Tianjian et al. (2011)	China	–	No	Smart transportation, 'dematerialization', smart work and smart appliances
Matsuoka (2011)	Japan	–	Yes	Green ICT Green materials Governance by green
Odadzic et al. (2012)	European Union (EU)	–	Yes	Hybrid smart system for energy control
Shuja et al. (2012)	22 datacenters	–	Yes	Message queue preemption at switches and wireless transmission scheduling techniques
Dong et al. (2012)	–	LP optimization model REO-hop	Yes	Efficient electrical grids Local adaptation along with the implementation of hardware-based techniques, such as sleep cycle and rate adaptation
Basmadjian et al. (2012)	–	–	Yes	Datacenters
Al-Daraiseh et al. (2013)	Saudi Arabia	–	Yes	Smart energy management system
Makela and Luukkinen (2013)	–	–	Yes	Incentives for companies that provide green services
Moura et al. (2013)	European Union	–	Yes	Smart grids Electric car development

(continued)

Table 9.1 (continued)

Author	Country/Region	Model/Method	Does ICT emit CO_2?	Suggestions
Bracco et al. (2013)	Italy (University of Genoa—The Savona Campus)	–	Yes	New lines and substations
Bracco et al. (2014)	Italy (University of Genoa—The Savona Campus)	Mathematical model	Yes	Smart grids/microgrid testbed
Aamir et al. (2014)	–	–	Yes	Use of green smart grid technology
Mahmod et al. (2014)	European Union	Amitran	–	Smart transportation Driver assistance systems Traffic management solutions Eco-driving support systems
Lee and Brahmasrene (2014)	Nine members from the Association of Southeast Asian Nations	Co-integration techniques Co-integrating regression estimation methods	Yes	The green use of ICT Innovative business models Progressive policy-making
Addis et al. (2014)	–	Mixed integer linear programming (MILP) model	Yes	Cloud services integration
Wu et al. (2014)	–	ClouSim	Yes	Efficient scheduling to better allocate resources and reduce working hours
Wang et al. (2015)	The UK	–	No	WMS, TMS, STMS, RFID, GPRS, web technology, smart vehicle systems, ERP, SCM
Kern et al. (2015)	–	Continuous energy efficiency measurement	Yes	Calculate the impact of ICTs and their production on climate change Calculation of carbon footprints (carbon footprint—CF) Life cycle assessment (LCA)
Caird et al. (2015)	The UK	–	Yes	Distance learning models

(continued)

Table 9.1 (continued)

Author	Country/Region	Model/Method	Does ICT emit CO_2?	Suggestions
Herrador et al. (2015)	European Union	'ISUMO' (Incentivized Sustainable Mobility)	No	Services that combine ICT transportation, a QR-based charger (quick response) and electric vehicles
Weiser et al. (2016)	Germany	–	Yes	Location awareness
Salahuddin et al. (2016)	Australia	–	Yes	Green ICT for green
Avgerinou, Bertoldi and Castellazzi (2017)	European Union	–	Yes	Code of conduct for data centre energy efficiency programme
Kanwal et al. (2017)	–	REHO scheme	Yes	Deactivation of base stations (Basestation) during peak hours Load balancing Operator aggregation Bandwidth expansion
Asongu et al. (2017)	44 Sub-Saharan African countries	–	No	Digital inclusion Access to affordable ICT Suitable ICT infrastructure
Perez-Prada et al. (2017)	Madrid, Spain	–	Yes	Green navigation systems (GNSs) Ecological route systems
Higón et al. (2017)	142 economies, split into 116 developing and 26 developed countries	–	Yes (developing countries) No (developed countries)	ICT development Changes in the structure of the economy (manufacturing to services)
Majeed (2018)	132 developed and developing economies	Pooled ordinary least squares (POLS) and generalized method of moments (GMM) estimation techniques	Yes	'Greening through ICTs' 'Greening of ICT'

(continued)

Table 9.1 (continued)

Author	Country/Region	Model/Method	Does ICT emit CO_2?	Suggestions
Park et al. (2018)	Austria, Belgium, Bulgaria, Croatia, Cyprus, Czech Republic, Denmark, Finland, France, Germany, Greece, Hungry, Ireland, Italy, Luxembourg, the Netherland, Poland, Portugal, Romania, Slovenia, Spain, Sweden and the UK	Panel data estimation approach, PMG, FMOLS, DOLS	Yes	Modern ICT devices Promotion of green sectors Alternative business models Fees application
Khan et al. (2018)	N-11 countries	Panel mean group (MG) Augmented mean group (AMG) estimation methods	Yes/No	Smart grid (home, transportation and services) Investment in internet innovations and R&D in the financial sector Credit to green ICT technology Government investmentssmart cities and green cities
Asongu (2018)	44 Sub-Saharan African countries	Generalized method of moments (GMM)	No	Reducing globalization-related information asymmetry Information sharing offices (public credit registries and private credit bureaus)
Asongu et al. (2018)	44 Sub-Saharan African countries	Generalized method of moments (GMM) Shapiro–Wilk normality test	Yes/No	Consolidation of ICTs

(continued)

Table 9.1 (continued)

Author	Country/Region	Model/Method	Does ICT emit CO_2?	Suggestions
Shabani and Shahnazi (2019)	Iran	Dynamic ordinary least squares (DOLS) Estimator Granger causality Environmental Kuznets curve (EKC)	Yes/No	Dematerialization
Akande et al. (2019)	129 cities in Europe	Pearson's linear correlation coefficient (data analysis)	Yes	Offer services such as education, health, mail, banking and commerce remotely for small cities by a broadband internet access
Aldakhil et al. (2019)	South Asia	–	Yes	ICT applications used in diversified economic restructuring
Chen et al. (2019)	China	Quantile regression	Yes/No	Use of ICT
Faisal et al. (2020)	Fast-emerging countries (FEC-BRICS) (except Russia)	Breusch and Pagan (1980) and Pesaran (2021) CD test	Yes	Environmentally friendly ICT equipment Easy access to credit
Khan et al. (2020)	91 countries	Pooled ordinary least squares Fixed-effects Model System-generalized method of moments Estimation techniques	No	ICT investments in developing countries Economic policies (change from the manufacturing sector to services sector) Develop alternative sources of energy

(continued)

Table 9.1 (continued)

Author	Country/Region	Model/Method	Does ICT emit CO_2?	Suggestions
Ulucak et al. (2020)	BRICS	Environmental Kuznets Curves (EKC)	No	Promote online shopping Development of e-conference culture Digitalization of economy Charge high fees for polluting companies Give subsidies to companies to invest in clean technologies Control the transfer of e-waste Carbon pricing and dumping Duties on e-waste
Chazarra-Zapata et al. (2020)	Murcia region (Southeastern Spain)	Life cycle assessment (LCA)	No	Integrate agriculture into sustainable technological development Use of renewable energies
Charfeddine and Kahia (2021)	Middle East and North Africa region (MENA)	Panel vector autoregressive model	Yes	Guarantee the required institutional capacity Provide energy from renewables Development of a suitable market structure to provide resources and opportunities to all companies
Ahmed et al. (2021)	Latin American and Caribbean	Panel econometric techniques	No/Yes	Investment in ICT infrastructure Subsidies and tax incentives Legislation and Regulations
Magazzino et al. (2021)	25 OECD countries	Dumitrescu–Hurlin panel	Yes	Legal frameworks that give more decision-making power to municipalities to decide on the use of ICT

Source Own elaboration (2021)

In the 1990s, with the expansion of the use of ICT, the member countries of the Organization for Economic Co-operation and Development (OECD) decided to constitute a study in 1997 called the Working Party on Indicators for the Information Society (WPIIS). The following subjects were addressed in this study: a definition of the ICT sector, broad and narrow definitions of e-commerce transactions, survey models of ICT use in businesses and households/individuals, classification of ICT products, measurement of investments in ICT, the guide to measurement in the Information Society, partnerships for this measurement and future challenges for the WPIIS (OECD 2007). Despite this effort, the indicators are still offered at a macro level, requiring adaptation to the specificities of the economies of developed and developing countries (Akande et al. 2019). The idea is to offer a proposal of indicators for measuring the impacts of ICT on the sustainability of a country.

One of the main advantages of ICT is its role in promoting interaction between countries and people, leading to a greater volume of exports and imports via electronic commerce in a globalized world economy (Faisal et al. 2020; Ulucak et al. 2020, p. 1). In addition, ICT, especially mobile broadband, is an essential platform for SDGs and an accelerator technology that helps societies go beyond the Business-As-Usual management trajectory (BAU). Developments in ICTs, including the Internet of Things (IoT), advanced robotics, artificial intelligence (AI), Big Data, 3D printing and others, will provide many tools for fundamental improvements in diverse sectors, such as health, education, energy and environmental protection (Kostoska and Kocarev 2019, p. 2).

However, fast-emerging countries own the infrastructure for this development, which makes them the largest emitters of CO_2 among emerging countries using ICT (Faisal et al. 2020). According to Akande et al. (2019), the non-sustainable use of resources to satisfy the immediate needs of the human species results in the increase of CO_2 in the atmosphere, among other impacts.

For Dakhal (2011), environmental governance benefits from social capital mediated by ICT, supporting organizations (ECOs) in inter- and intra-organizational interactions. ICTs can also help in the detection of companies with problems related to terrorism and money laundering; to this end, information networks build investment exclusion lists to protect countries and institutions from companies with ESG-related problems (Hisano et al. 2020).

Therefore, Pan et al. (2021) and Pan and Zhang (2020) note that research related to digital sustainability can help countries in solving the climate crisis, achieving Sustainable Development Goals through artificial intelligence (AI), machine learning (ML) and the Internet of Things (IoT). Pan and Zhang (2020) state that digital sustainability is the convergence of digital imperatives and sustainability that involves a transdisciplinary approach to deploying digital technologies in addressing sustainability issues. Given its digital enablement nature, digital sustainability could enable ICT researchers to play a key role in the trajectory towards achieving the SDG goals.

The ICTs that help in the process of bringing the operations of various sectors (including finance) closer to sustainability goals can be Distributed Ledger Technologies (DLT), Big Data or AI. DLTs can assist companies and countries in ESG with creating transparent, agile and automated data collection processes. In a DLT network, all participants are privileged and equitably connected, avoiding the need for an administrative authority that centralizes all operations. This form of organization allows control to be distributed among several nodes, avoiding points of failure in the system. Thus, information is transmitted to all sectors in an integrated and tracked manner. This avoids deviations and misunderstandings in companies and supply chains where different companies and institutions interact. This better relationship through ICT allows initiatives such as ESG to be carried out together in a transparent way (Dario et al. 2021).

In recent years, policies have been recommended to curb greenhouse gases and CO_2 emissions by improving energy efficiency, conserving energy and planning energy use strategies. These policies have been motivated mainly by the high level of CO_2 emission, as a consequence of the use of non-renewable energy sources, and also the high level of participation of non-renewable energies in the energy matrix. Currently, these non-renewable sources of energy represent more than 80% of global energy consumption. In particular, two important policies have been widely discussed, due to their high potential for avoiding increased CO_2 emissions: (1) the promotion of renewable energy and (2) the expansion of the use of ICT (Charfeddine and Kahia 2021).

At the last Climate Conference (UN COP 26 2021), several countries committed to decreasing environmental impacts, such as reducing deforestation by 2030 (100 countries, including Brazil), reducing methane emissions by 30% by 2030 and eliminating the use of coal as a source of energy. The use of technologies and ICT was one of the points with many mitigation proposals by governments and society around the world, so that the necessary steps are taken in the transition to net zero (Dwivedi et al. 2022).

9.4 CO_2 and ICT Emission Cases

Box 9.4.1: Telecommunications Industry—Yang et al. (2020)

Five telecommunications' companies located in Asia (3 in China and 2 in Singapore—from Telco1 to Telco5) have adopted environmental strategies aligned with ICT. Telco1 has a green data centre (new servers, routers, switches and new methods of cooling), Telco2 has a green data centre (cloud computing [EMC storage], blade servers, new design of transformers), Telco3 has a green data centre (upgraded servers and cooling facilities, small-scale adoption of cloud computing), Telco4 has a green data centre and Telco 5 has solar-powered mobile base stations, an environmentally friendly office building with low power consumption.

By dividing the analysis into five drivers—green IT initiatives, cost reduction, market demand, government pressure, corporate social responsibilities (CSR) and Reconciliation of 'doing good' and 'doing well'—a ranking division was performed, according to H (high), M (medium) and L (low). Among all the five companies, Telco5 was the one that stood out the most, with four 'H' out of the six drivers.

Regarding qualitative analyses, 'green IT initiatives' were not considered by companies as mere technical practices, but also as an essential strategy, and some of them as key tools to help organizations to achieve operational excellence and corporate sustainability. Some particular sustainability-related objectives were achieved through these initiatives, such as reducing energy consumption (solar power station in Telco5) and controlling ICT costs at the individual, departmental and organizational levels (Telco2). Furthermore, economic results are more effective when 'green IT initiatives' are aligned with overall business strategies than when they are not.

As for the 'market demand' driver, the results showed that for better or more sustainable products, reconciliation with the 'cost reduction' objectives is less feasible. On the other hand, 'cost reduction' aligns better with the 'government pressure' driver. The latter competes more with economic development drivers ('cost reduction' and 'corporate social responsibility') when it comes to state-owned companies (Telco1, Telco2 and Telco3, in China).

However, except for the relatively low 'government pressure' in the two Singapore companies, the other drivers showed no differences between the two countries. There was an interesting finding in this driver, since it and SSR are mandatory and not just restrictions, because the Chinese government prefers a reward system, whereas the Singapore government tends towards a punishment system.

The results confirmed that legitimacy (reflected by 'government pressure' as a driver of authority) and green responsibility (indicated by 'CSR' as a driver of morality) influence 'green IT initiatives'. While the Chinese government provides a research fund for organizations to conduct 'green IT initiatives', especially for state-owned enterprises (Telco 1 to 3), in contrast, the Singapore government has strict rules on corporate sustainability, which forces companies to experiment with green IT applications. CSR is mandatory for large companies and they are required to discuss their environmental sustainability practices in their annual reports. Green IT comes from an important chapter of CSR for most companies in the IT industry. These analyses show that ecological legitimacy and responsibility have significant impacts on green IT initiatives.

'Doing good' (environmental sustainability) and 'Doing well' (business profitability) are reconciled when organizations consider short-term investments and long-term benefits, appropriate design and strategies to withstand intense external pressure.

Box 9.4.2: Suzano (Ondei 2021; Schalka 2021)

Suzano is a publicly traded company in the pulp and paper sector, whose total capital is represented by 1361 million shares, of which about 46% are free float. The company has ambitious goals beyond processing forests, which includes removing 10 million tonnes of plastics from its production system by 2030. After the 26th United Nations Conference on Climate Change (COP 26), Suzano's environmental focus includes goals, such as removing 40 million tCO_2e in the period from 2020 to 2025, reducing the intensity of carbon emissions (Scopes 1 and 2) per tonne produced (tCO_2eq/t of product) by 15% by 2025 and offering 10 million tonnes of products from renewable sources to replace materials produced from fossil sources. Among the various innovation fronts, it is important to point out the development of textile fibres from biomass, the production of lignin for application in resins and the development of transportation biofuels from wood. In addition, in 2021 Suzano announced another long-term goal that will make a great contribution to the planet: the company will connect half a million hectares of priority areas for the conservation of Biodiversity in the Cerrado, the Atlantic Forest and the Amazon by 2030. As for joining the carbon market, to offset the emission the company joined the Climate Action 100 + initiative, led by investors to ensure that the largest corporate emitters of greenhouse gases adopt the necessary measures on climate change, and started to incorporate the Assessing Low-Carbon Transition (ACT) initiative for the pulp and paper sector. It has also recently joined the 'Science Based Target Initiative' (SBTi) and the 'Business Ambition for 1.5 °C' campaign and the 'Race to Zero'.

Two goals are considered an internal responsibility: generating income for non-employees and investing in diversity. Regarding income generation, the company gives as an example the Beehives Program, which produces honey in eucalyptus cultivation areas. One of the hubs is in the municipality of Mucuri (BA), where 77 beekeepers harvested 15.4 tonnes of honey between January and April with technical support from the company. Regarding the issue of the inclusion of minorities, the company has been structuring greater participation of women and Black people. The company expects by 2025 to have 30% of leadership positions occupied by women and Black people. In the case of women, the programme has evolved considerably, but there are some hindrances in the case of Black people, due to a historical problem of educational training. Some solutions would be changing the search profile and the identification and qualification of talents.

Box 9.4.3: Renner (Ondei 2021; Renner 2021)

The year 2021 was the deadline set by Renner (located in Brazil) to achieve a series of sustainability commitments, announced publicly in 2018. The agenda includes reaching 80% of clothing products made with raw materials and processes with less impacting—exclusively using certified cotton; supplying 75% of corporate energy consumption (stores, distribution centres and offices) with low-impact renewable sources; having 100% of the cotton used with socio-environmental certification (and other suppliers); and reducing CO_2 emissions by 20% when compared to 2017—which have since been neutralized in compensation programmes. In September 2021, Renner acquired the start-up Repassa, which bets on the Circular Economy of clothes, focusing on the female audience of classes B and C and selling used clothes in a thrift store in a large shopping centre in the city of Campinas and online.

Some numbers already show this acceleration of ESG practices, as the company reached a total of 130 million pieces with some less-impacting raw materials, of which 75 million were in 2020, according to Eduardo Ferlauto, general manager of sustainability. That same year, corporate consumption of energy based on low-impact renewable sources reached 65% of the total. The company has the Re Seal—Responsible Fashion, which identifies sustainability initiatives: from products made with raw materials and/or processes with low impact on the planet to Ecostyle, the reverse logistics programme that disposes in an environmentally correct manner perfumery packaging and bottles and clothing items.

CO_2 emissions fell by 36.75%, exceeding the target, driven in part by the change in the energy standard and by restructuring in the chain, and in part by the reduction of operations. R$ 5.1 million were directly invested in actions to support vulnerable people and health professionals and institutions on the front lines of the fight against Covid-19. There were also efforts to support the supply chain, with a donation of R$ 1.5 million to small companies in a critical situation and the intermediation of R$ 87 million in loans from BNDES.

All ESG initiatives are based on four pillars: eco-efficient management; sustainable products and services; responsible suppliers; and engagement of employees, communities and customers. Sometimes that means giving up immediate results, as in the case of Re Jeans—made from textile leftovers, which would go to landfills and are now shredded and returned to the production cycle. The idea is to increase efficiency, but first it was necessary to invest in research, equipment, and supplier training.

Renner is in first place in the ESG index launched by B3 and S&P Dow Jones in 2020. It was also included for seven consecutive years in the ISE B3 and for the sixth year in the Dow Jones Sustainability (DJSI)—in this case, one of the six companies in the fashion retail world.

9.5 Indication of How the Conceptual Framework Will Be Incorporated into the Process of Development, Implementation, and Monitoring of the Strategy and Production Organization

Based on the literature, some suggestions for reducing CO_2 emissions and energy consumption are presented, based on the sources presented in session 2 to reduce CO_2 emission and to lower costs:

- Implement a building management system in which all subsystems (lighting, air conditioning, electrical, and fire prevention systems) are integrated in order to increase employee efficiency.
- Place screens with information on energy consumption per floor.
- Exchange of less efficient lamps with new ones to replace two with one.
- Plan buildings with spaces privileging natural light (elevators and corridors); lamps with longer life cycle installed in common areas of the building (Miyashita et al. 2010).
- Design windows to increase ventilation and maintain room temperature in order to reduce the use of air conditioning.
- Use durable paint on the external parts of the building in order to extend the maintenance cycle.
- Reuse rainwater and kitchen water to clean bathrooms.
- Heat water at night when the local power rate is lower.
- Press non-organic waste.
- Compost waste from food and beverages (cafeterias, restaurants, meal stations, etc.). (Miyashita et al. 2010).

Regarding the use of ICT to make practices more friendly to the environment:

- Reduce paper consumption.
- Hold teleconferences for the company and videoconferences for distant meetings, reducing travelling and consequently CO_2 emissions from transportation (Tianjian et al. 2011; Ulucak et al. 2020).
- Improve mobility and security (Miyashita et al. 2010).
- Use smart transportation (Tianjian et al. 2011).
- Use smart work and smart appliances (Tianjian et al. 2011).
- Implement dematerialization (Tianjian et al. 2011; Ulucak et al. 2020).
- Order meals by phone, rather than cooking at home or buying online (Tianjian et al. 2011; Ulucak et al. 2020).

The following renewable energy sources could help companies to reduce energy consumption:

- Fuel cells.
- Photovoltaic cell.
- Wind turbine generators, micro hydro-generators.

- Fresh air cooling.
- Hybrid intelligent system for energy control, combining renewable energy and other innovative energy use solutions with traditional sources (Odadzic et al. 2012; Moura et al. 2013).

Transportation emits high levels of CO_2 and greenhouse gases. In this case, smart transportation would be a solution to reduce these impacts, through the application of several well-known ICTs in transportation (cargo), storage, and warehouse activities: Smart Transportation Systems (ITS) (Mahmod et al. 2014); Warehouse Management Systems (WMS), Transportation Management Systems (TMS), Smart Transportation Management Systems (STMS) e Radio Frequency Identification (RFID), and Enterprise Resource Planning (ERP) (Wang et al. 2015). In this logistic activity, in small areas, Urban Distribution Centres (UDC) are great structures to prevent damage to the structure of historic cities (Carvalho et al. 2019a; de Carvalho et al. 2019a, b). Likewise, the Green Navigation Systems (GNSs) have a great potential to reduce traffic consumption and emissions from passenger transportation. Since conscious drivers (green drivers) tend to choose the shortest routes within downtown areas, ecological route systems can become effective tools to eliminate the effects of climate change (Perez-Prada et al. 2017; Garcia-Castro et al. 2017).

Cloud computing can be used optimally, helping to schedule jobs in order to allocate resources efficiently, customizing the service according to customer needs (Wu et al. 2014).

To reduce global warming by mobile communication, base stations (Basestation) may be deactivated during peak hours, and load balancing, carrier aggregation, and bandwidth expansion can be performed (Kanwal et al. 2017).

Efficient use of electricity and modern ICT devices (such as smart grids—SG), alternative business models, and tariff applications would not only reduce energy consumption, but also control CO_2 emissions and e-waste (Park et al. 2018; Khan et al. 2018).

In addition to innovative strategies and initiatives involving technologies and strategies to reduce electricity consumption in companies from different sectors and college education institutions, political and economic strategies from governments could advance in the search for:

- Easy access to credit in order to promote the purchase and use of environmentally friendly ICT equipment (Faisal et al. 2020).
- Investment in ICT infrastructure.
- Promoting the role of the internet to revolutionize online shopping, money transfer, online banking, traffic control, and distance and remote education (tax incentives) (Magazzino et al. 2021).
- Laws and regulations for ICT to safeguard the interests of customers and vendors and gain the trust of all stakeholders (Ahmed et al. 2021).

9.6 Summary of Main Ideas and Concepts

- Governance—governance captures the change from a traditional government structure towards a mechanism in which the state values private sector (market) and third sector (civil society) actions in the decision-making process.
- Information and Communication Technologies (ICT)—hardware, software, telecommunications, automation, multimedia resources, data organization resources, information systems, services, businesses, users, and complex relationships involved in collecting, using, analysing, and using information (Gomes and Ribeiro 2014, p. 127).
- Dematerialization—the way of delivering services and replacing physical interactions and products.
- Greenhouse gases (GhGs)—the gases internationally recognized as greenhouse gases regulated by the Kyoto Protocol are the following: carbon dioxide (CO_2), methane (CH_4), nitrous oxide (N_2O), sulphur hexafluoride (SF_6), and two families of gases, hydrofluorocarbon (HFC) and perfluorocarbon (PFC) (ABNT 2021).
- Fossil fuels—the top 3 are coal, oil, and natural gas.
- Carbon credit/carbon footprint—carbon credits are generated by countries that have already met their greenhouse gas emission reduction targets (Junqueira 2021).
- ICTs affect energy consumption and CO_2 emissions in three ways. The first is by usage (production of software and hardware), the second is the substitution effect (reorganization of the production process), and the third is related to costs, the increase in demand for other goods and services arising from falling prices.
- ICTs can lead to advantages in terms of operational cost reduction, reductions in energy consumption and CO_2 emissions with the maintenance of systems and technology security, the reduction of travel, the management of energy reduction systems and transportation, using of electric cars, improving the management of energy consumption by the mobile communication industry, and reducing the use of paper for printing.
- The biggest challenge of cloud computing is to reduce its carbon footprint. The proposal for scheduling work, in order to allocate resources efficiently, is to customize the service to customer's needs and connect service centres and communication networks, managing energy expenditure, optimizing costs, and reducing the emission of CO_2.
- The online teaching model achieved the lowest energy consumption and carbon emissions, although there was a rebound effect associated with increased energy and paper consumption for prints related to the use of ICT (Caird et al. 2015).
- While developing nations still need to invest in ICT, developed nations have already reached the required level of these technologies. And in developed nations, CO_2 emissions are reduced as ICTs reach a higher level of development (Higón et al. 2017).
- 'Greening through ICT'—achieving de-carbonization in another sector through ICT services.

- 'Greening of ICT'—means carbon reduction in the ICT sector.
- The use of smart grids to reduce CO_2 emissions, such as smart devices for homes, the smart renewable and smart electricity services.
- Measuring environmental degradation is important, including the impact of ICT on the environmental dimension of ESG by CO_2 emissions per capita and liquid fuel.
- Among the environmental risks, CO_2 is the biggest contributor to global warming compared to other greenhouse gases (greenhouse gases).
- In some studies, ICTs have led to a 40% reduction in CO_2 emissions (compared to 1990 levels), an increase in renewable energy sharing by 27%, and an improvement in energy efficiency by at least 27%. In addition, compliance with the requirements of the objectives of some SDGs on the use of 'reclaimed water', renewable energies, efficient use of natural resources, efficient use of land to avoid abandonment, and consequent desertification (SDG 6, 7, 12 e 15) (Chazarra-Zapata et al. 2020).
- There is no consensus in the literature about the relationship between ICT and CO_2 emission.

9.7 Questions for Reflection and Study

(1) What are the advantages of adopting ICT in a company's operations in terms of CO_2 emissions?
(2) What are the disadvantages of adopting ICT in a company's operations in terms of CO_2 emissions?
(3) Summarize the strategies presented by the authors to reduce CO_2 emissions.
(4) How could the impact of using mobile communication technology on energy consumption be minimized?

9.8 Cases to Be Solved

(1) Imagine that employees in the administrative and operational divisions (including drivers) of a company in the logistics industry (cargo transportation services) have company cell phones, and the company plans to reduce energy consumption and CO_2 emissions. What initiatives can this company carry out?
(2) A Higher Education Institution (HEI) has a high level of energy consumption and CO_2 emissions, since its campus is composed of several buildings equipped with air conditioning, several ICT, projection equipment, computers and printers. What could this HEI do to reduce energy consumption and CO_2 emissions?

References

Aamir M, Uqaili MA, Chowdhry BS, Rub R, Ali A, Siddiqui M (2014) Framework for impact analysis of green smart grid deployment in energy deficit countries. In: International multi topic conference. Springer, Cham, pp 1–12. https://doi.org/10.1007/978-3-319-10987-9_1

ABNT. O que são gases do efeito estufa? Available at https://www.abntonline.com.br/sustentabilidade/GHG/O_que_%C3%A9_gee Accessed Dec 2021

Addis B, Ardagna D, Capone A, Carello G (2014) Energy-aware joint management of networks and cloud infrastructures. Comput Netw 70:75–95. https://doi.org/10.1016/j.comnet.2014.04.011

Ahmed Z, Nathaniel SP, Shahbaz M (2021) The criticality of information and communication technology and human capital in environmental sustainability: evidence from Latin American and Caribbean countries. J Clean Prod 286:125529. https://doi.org/10.1016/j.jclepro.2020.125529

Aich S, Thakur A, Nanda D, Tripathy S, Kim HC (2021) Factors affecting ESG towards impact on investment: a structural approach. Sustainability 13(19):10868. https://doi.org/10.3390/su131910868

Akande A, Cabral P, Casteleyn S (2019) Assessing the gap between technology and the environmental sustainability of European cities. Inf Syst Front 21(3):581–604. https://doi.org/10.1007/s10796-019-09903-3

Alataş S (2021) The role of information and communication technologies for environmental sustainability: evidence from a large panel data analysis. J Environ Manage 293:112889. https://doi.org/10.1016/j.jenvman.2021.112889

Aldakhil AM, Zaheer A, Younas S, Nassani AA, Abro MMQ, Zaman K (2019) Efficiently managing green information and communication technologies, high-technology exports, and research and development expenditures: a case study. J Clean Prod 240:118164. https://doi.org/10.1016/j.jclepro.2019.118164

Al-Daraiseh A, Shah N, El-Qawasmeh E (2013) An intelligent energy management system for educational buildings. Int J Distrib Sens Netw 9(9):209803. https://doi.org/10.1155/2013/209803

Asongu SA (2018) ICT, openness and CO_2 emissions in Africa. Environ Sci Pollut Res 25(10):9351–9359. https://doi.org/10.1007/s11356-018-1239-4

Asongu SA, Le Roux S, Biekpe N (2017) Environmental degradation, ICT and inclusive development in Sub-Saharan Africa. Energy Policy 111:353–361. https://doi.org/10.1016/j.enpol.2017.09.049

Asongu SA, Le Roux S, Biekpe N (2018) Enhancing ICT for environmental sustainability in sub-Saharan Africa. Technol Forecast Soc Chang 127:209–216. https://doi.org/10.1016/j.techfore.2017.09.022

Avgerinou M, Bertoldi P, Castellazzi L (2017) Trends in data centre energy consumption under the European code of conduct for data centre energy efficiency. Energies 10(10):1470. https://doi.org/10.3390/en10101470

Basmadjian R, De Meer H, Lent R, Giuliani G (2012) Cloud computing and its interest in saving energy: the use case of a private cloud. J Cloud Comput: Adv, Syst Appl 1(1):1–25. https://doi.org/10.1186/2192-113X-1-5

Bracco S, Delfino F, Pampararo F, Robba M, Rossi M (2014) A mathematical model for the optimal operation of the University of Genoa smart polygeneration microgrid: evaluation of technical, economic and environmental performance indicators. Energy 64:912–922. https://doi.org/10.1016/j.energy.2013.10.039

Bracco S, Delfino F, Pampararo F, Robba M, Rossi M (2013) The University of Genoa smart polygeneration microgrid test-bed facility: the overall system, the technologies and the research challenges. Renew Sustain Energy 18:442e59

Breusch TS, Pagan AR (1980) The Lagrange multiplier test and its applications to model specification in econometrics. Rev Econ Stud 47(1):239–253. https://doi.org/10.2307/2297111

Caird S, Lane A, Swithenby E, Roy R, Potter S (2015) Design of higher education teaching models and carbon impacts. Int J Sustain High Educ 16(1):96–111. https://doi.org/10.1108/IJSHE-06-2013-0065

Carvalho NLA, Ribeiro PCC, Oliveira LK, Silva JEAR, Vieira JGV (2019a) Criteria to implement UDCs in historical cities: a Brazilian case study. Eur Transp Trasporti Europei 72(1):1–29

Charfeddine L, Kahia M (2021) Do information and communication technology and renewable energy use matter for carbon dioxide emissions reduction? Evidence from the Middle East and North Africa region. J Clean Prod 327:129410. https://doi.org/10.1016/j.jclepro.2021.129410

Chazarra-Zapata J, Parras-Burgos D, Pérez-de-la-Cruz FJ, Ruíz-Canales A, Molina-Martínez JM (2020) Reducing the carbon footprint of the water-energy binomial through governance and ICT. A case study. Water 12(11):3187. https://doi.org/10.3390/w12113187

Chen X, Gong X, Li D, Zhang J (2019) Can information and communication technology reduce CO_2 emission? A quantile regression analysis. Environ Sci Pollut Res 26(32):32977–32992. https://doi.org/10.1007/s11356-019-06380-8

Dario C, Sabrina L, Landriault E, De Vega P (2021) DLT to boost efficiency for financial intermediaries. An application in ESG reporting activities. Technol Anal Strateg Manag 1–14. https://doi.org/10.1080/09537325.2021.1999921

de Carvalho NLA, Ribeiro PCC, García-Martos C, Fernández CG, Vieira JGV (2019b) Urban distribution centres in historical cities from the perspective of residents, retailers and carriers. Res Transp Econ 77:100744

Dhakal SP (2011) Can environmental governance benefit from an ICT-social capital nexus in civil society?. TripleC: Commun, Capitalism Critique. Open Access J Global Sustain Inf Soc 9(2):551–565. https://doi.org/10.31269/triplec.v9i2.260

Dong X, El-Gorashi T, Elmirghani JM (2012) Energy efficient optical networks with minimized non-renewable power consumption. J Networks 7(5):821

Dwivedi YK, Hughes L, Kar AK, Baabdullah AM, Grover P, Abbas R et al (2022) Climate change and COP26: are digital technologies and information management part of the problem or the solution? An editorial reflection and call to action. Int J Inf Manage 63:102456. https://doi.org/10.1016/j.ijinfomgt.2021.102456

Elkington J (1998) Partnerships from cannibals with forks: the triple bottom line of 21st-century business. Environ Qual Manage 8(1):37–51. https://doi.org/10.1002/tqem.3310080106

Faisal F, Tursoy T, Pervaiz R (2020) Does ICT lessen CO_2 emissions for fast-emerging economies? An application of the heterogeneous panel estimations. Environ Sci Pollut Res 1–12. https://doi.org/10.1007/s11356-019-07582-w

Garcia-Castro A, Monzon A, Valdes C, Romana M (2017) Modeling different penetration rates of eco-driving in urban areas: impacts on traffic flow and emissions. Int J Sustain Transp 11(4):282–294. https://doi.org/10.1080/15568318.2016.1252972

Gomes CFS, Ribeiro PCC (2014) Gestão da Cadeia de Suprimentos integrada à Tecnologia da Informação. Cengage Learning, SP

Herrador M, Carvalho A, Feito FR (2015) An incentive-based solution of sustainable mobility for economic growth and CO_2 emissions reduction. Sustainability 7(5):6119–6148. https://doi.org/10.3390/su7056119

Higón DA, Gholami R, Shirazi F (2017) ICT and environmental sustainability: a global perspective. Telematics Inform 34(4):85–95. https://doi.org/10.1016/j.tele.2017.01.001

Hisano R, Sornette D, Mizuno T (2020) Prediction of ESG compliance using a heterogeneous information network. J Big Data 7(1):1–19. https://doi.org/10.1186/s40537-020-00295-9

Junqueira C (2021) Como o crédito de carbono pode ajudar o meio ambiente? ESG investimento. Exame invest. 26 Nov 2021. Available at https://exame.com/blog/cristina-junqueira/como-o-credito-de-carbono-pode-ajudar-o-meio-ambiente/. Accessed Dec 2021

Kanwal K, Safdar G, Ur-Rehman M (2017) Energy efficiency led reduced CO_2 emission in green LTE networks. Ozp -Institut fur Staats und Politikwissenschaft 4(14):e2. https://eudl.eu/doi/10.4108/eai.4-10-2017.153160

Kern E, Dick M, Naumann S, Hiller T (2015) Impacts of software and its engineering on the carbon footprint of ICT. Environ Impact Assess Rev 52:53–61. https://doi.org/10.1016/j.eiar.2014.07.003

Khan D, Khan N, Baloch MA, Saud S, Fatima T (2018) The effect of ICT on CO_2 emissions in emerging economies: does the level of income matters? Environ Sci Pollut Res 25(23):22850–22860. https://doi.org/10.1007/s11356-018-2379-2

Khan FN, Sana A, Arif U (2020) Information and communication technology (ICT) and environmental sustainability: a panel data analysis. Environ Sci Pollut Res 27(29):36718–36731. https://doi.org/10.1007/s11356-020-09704-1

Kostoska O, Kocarev L (2019) A novel ICT framework for sustainable development goals. Sustainability 11(7):1961. https://doi.org/10.3390/su11071961

Lee JW, Brahmasrene T (2014) ICT, CO_2 emissions and economic growth: evidence from a panel of ASEAN. Glob Econ Rev 43(2):93–109. https://doi.org/10.1080/1226508X.2014.917803

Magazzino C, Mele M, Morelli G, Schneider N (2021) The nexus between information technology and environmental pollution: application of a new machine learning algorithm to OECD countries. Utilities Policy 72:101256. https://doi.org/10.1016/j.jup.2021.101256

Mahmod MKM, Jonkers E, Klunder G, Benz T, Andrew W (2014) The Amitran methodology framework for evaluating the impact of ICT-based measures on CO_2 emissions in the transport field. IET Intell Transp Syst 1–11. https://doi.org/10.1049/iet-its.2014.0058

Majeed MT (2018) Information and communication technology (ICT) and environmental sustainability in developed and developing countries. Pak J Commer Soc Sci 12(3):758–783

Makela T, Luukkainen S (2013) Incentives to apply green cloud computing. J Theor Appl Electron Commer Res 8(3):74–86. https://doi.org/10.4067/S0718-18762013000300006

Matsuoka M (2011) Environment and energy technologies toward a green society. NTT Techn Rev 9(2):1–6

Miyashita T, Sakurai M, Tanaka S, Komura T (2010) Total ecoslsoution for office—a state-of-the-art office model that achieves a having CO_2 emissions and implement innovative work style. NEC Tech J 5(4):23–27

Moura PS, López GL, Moreno JI, De Almeida AT (2013) The role of smart grids to foster energy efficiency. Energ Effi 6(4):621–639. https://doi.org/10.1007/s12053-013-9205-y

Odadzic B, Lukic NM, Jankovic M (2012) The application of femtocells as a technical solution for a telecomunication provider. TEM J 3

Organisation for Economic Co-operation and Development (OECD) (2007) Working party on indicators for the information society. DSTI/ICCP/IIS (2006)2/FINAL. Directorate for science, technology and industry committee for information, computer and communications policy. 05 March 2007. JT03223017. Available at https://www.oecd.org/sti/38217340.pdf. Accessed Dec 2021

Ondei V (2021) Especial ESG: as 10 empresas que se destacam na adoção de práticas sustentáveis. Forbes. Forbes ESG, edição 87, lançada em maio de 2021. Atualizada em 22 de agosto de 2021. Available at https://forbes.com.br/forbesesg/2021/08/especial-esg-as-10-empresas-que-se-destacam-na-adocao-de-praticas-sustentaveis/. Accessed Dec 2021

Pan SL, Carter L, Tim Y, Sandeep MS (2021) Digital sustainability, climate change, and information systems solutions: opportunities for future research. Int J Inf Manag 102444. https://doi.org/10.1016/j.ijinfomgt.2021.102444

Pan SL, Zhang S (2020) From fighting COVID-19 pandemic to tackling sustainable development goals: an opportunity for responsible information systems research. Int J Inf Manage 55:102196. https://doi.org/10.1016/j.ijinfomgt.2020.102196

Park Y, Meng F, Baloch MA (2018) The effect of ICT, financial development, growth, and trade openness on CO_2 emissions: an empirical analysis. Environ Sci Pollut Res 25(30):30708–30719. https://doi.org/10.1007/s11356-018-3108-6

Perez-Prada F, Monzon A, Valdes C (2017) Managing traffic flows for cleaner cities: the role of green navigation systems. Energies 10(6):791. https://doi.org/10.3390/en10060791

Pesaran MH (2021) General diagnostic tests for cross-sectional dependence in panels. Empir Econ 60(1):13–50. https://doi.org/10.1007/s00181-020-01875-7
Renner (2021) Re Moda Responsável. Sustentabilidade. Available at https://www.lojasrenner.com.br/sustentabilidade Accessed Dec 2021
Sachs I (1993) Estratégias de transição para o século XXI: desenvolvimento e meio ambiente. Editora Nobel, São Paulo
Salahuddin M, Alam K, Ozturk I (2016) Is rapid growth in internet usage environmentally sustainable for Australia? An empirical investigation. Environ Sci Pollut Res 23(5):4700–4713. https://doi.org/10.1007/s11356-015-5689-7
Schalka W (2021) Para a Suzano 2050 é agora. Posicionamento da Suzano sobre mudanças climáticas. Suzano. Available at https://15a10fbf-7067-4a73-81e1-7aeaa165214a.filesusr.com/ugd/423557_b5355a0a709445839332e49b556ed177.pdf. Accessed Dec 2021
Shabani ZD, Shahnazi R (2019) Energy consumption, carbon dioxide emissions, information and communications technology, and gross domestic product in Iranian economic sectors: a panel causality analysis. Energy 169:1064–1078. https://doi.org/10.1016/j.energy.2018.11.062
Shuja J, Madani SA, Bilal K, Hayat K, Khan SU, Sarwar S (2012) Energy-efficient data centers. Computing 94(12):973–994. https://doi.org/10.1007/s00607-012-0211-2
Tianjian Y, Yiwen H, Ping Z, Huaying S, Xi L (2011) Low-carbon telecom solution for China's emission reduction and future forecasts. China Commun 8(3):52–65
Ulucak R, Danish, Khan SUD (2020) Does information and communication technology affect CO2 mitigation under the pathway of sustainable development during the mode of globalization?. Sustainable Development 28(4):857–867. https://doi.org/10.1002/sd.2041
Wang Y, Rodrigues VS, Evans L (2015) The use of ICT in road freight transport for CO_2 reduction–an exploratory study of UK's grocery retail industry. Int J Logistics Manag. https://doi.org/10.1108/IJLM-02-2013-0021
Weiser P, Scheider S, Bucher D, Kiefer P, Raubal M (2016) Towards sustainable mobility behavior: research challenges for location-aware information and communication technology. GeoInformatica 20(2):213–239. https://doi.org/10.1007/s10707-015-0242-x
Wu CM, Chang RS, Chan HY (2014) A green energy-efficient scheduling algorithm using the DVFS technique for cloud datacenters. Futur Gener Comput Syst 37:141–147. https://doi.org/10.1016/j.future.2013.06.009
Yang X, Li Y, Kang L (2020) Reconciling doing good and doing well in organizations green IT initiatives: a multi-case analysis. Int J Inf Manage 51:102052. https://doi.org/10.1016/j.ijinfomgt.2019.102052

Priscilla Cristina Cabral Ribeiro is Associate Professor of the Production Engineering Department at Universidade Federal Fluminense (UFF), teaching economics to undergraduate majors and supply chain management to graduate majors. She obtained bachelor's degree in economics (1995) at Universidade Rural do Rio de Janeiro, master's (MSc) in production engineering (1998) at Universidade Federal do Rio de Janeiro, and doctor degree in production engineering (2009) at Universidade Federal de São Carlos. From 2008 to 2009, she was Visiting Scholar at Utah State University during her Ph.D. programme. She has been reviewing papers for journals such as Sustainability and Heliyon. She is studying sustainability, supply chain management, and artificial intelligence (AI), focusing on sustainable procurement, a green warehouse, suppliers' evaluation, and food waste. Her research has appeared in journals, such as the International Journal of Production Economics, the International Journal of Information Management, and Environmental Science and Pollution Research. She supervised around 80 students in research projects and undergraduate and graduate theses until now.

Chapter 10
Understanding Sustainable Operations Frameworks

Edson Pinheiro De Lima, Sergio E. Gouvea Da Costa, Carla Gonçalves Machado, and Pablo Carpejani

Abstract Sustainable operations practices have been intensively assessed in the last few years and are opportunities to test the developed theory. This chapter aims to describe and analyze some 'professional' frameworks that professional associations and consulting firms are proposing. Fundamental content analysis is developed to identify evidence for sustainable concept adoption, CMMI adherence, and scope definition. Results show an incomplete implementation focusing primarily on environmental aspects; however, evidence for strategic sustainability management, governance structures based on multiple stakeholders, and models' evolution based on continuous improvement and innovation management were found. Recently ESG frameworks are being revisited, circular economy models, and regenerative strategies.

Keywords Sustainability · Sustainable operations · Operations management · ESG

E. Pinheiro De Lima (✉) · S. E. Gouvea Da Costa
Industrial and Systems Engineering, Federal University of Technology-Parana (UTFPR), Parana, Brazil
e-mail: pinheiro@utfpr.edu.br

S. E. Gouvea Da Costa
e-mail: gouvea@utfpr.edu.br

C. G. Machado
Business Administration at Jönköping International Business School, Jönköping University, Jönköping, Sweden
e-mail: carla.goncalvesmachado@ju.se

P. Carpejani
Business of Pontifical Catholic University of Parana (PUCPR), Curitiba, Brazil
e-mail: pablocarpejani@hotmail.com

M. J. Meiriño and O. L. G. Quelhas (eds.), *Critical Factors Associated with Environmental, Social and Governance Management in Organizations*, Accounting, Finance, Sustainability, Governance & Fraud: Theory and Application, https://doi.org/10.1007/978-981-96-1725-8_10

10.1 Understanding the Evolution of Sustainability Frameworks

Sustainability is a research theme that has been intensively worked on in the OM community for the latest ten years, evolving to a more integrated and systemic view (Gunasekaran and Ngai 2012; Seuring and Muller 2008).

The evolution of sustainability thought at 'business schools' could be traced back by analyzing some of M. E. Porter's publications. A significant move in strategic thinking characterized the 80s and '90s; that is, structural and process-based frameworks were proposed for studying competitiveness from companies to nation's levels. Porter's (1979) structural five forces competitive model, Porter's (1987) generic strategies and value chain framework, and Porter's structural competitive 'diamond' framework (1990) are good examples of describing in a market-based approach how strategic theory evolved in this period. Remarkably, the main focus is defined in an economic sense, although a multivariable system of aspects defines competition patterns.

In 1991, Porter (1991) wrote an article in Scientific American Magazine that focused on discussing 'green' opportunities for American companies' competitiveness. Porter and van der Linde (1995a, 1995b) advocate that a resource productivity framework could be used to solve green and competitiveness 'stalemate'; companies could foresee environmental issues as opportunities for differentiation through innovation. The authors strongly believe that the orientation should shift from pollution control to resource productivity, and success must involve innovation-based solutions that promote both environmentalism and industrial competitiveness. Porter and Esty (1998) concluded that industrial ecology offers a promising tool for enhancing resource productivity through its systemic approach at the firm level. Industrial ecology can promote innovation and help managers find opportunities, both inside and outside their operations, approaching the entire set of companies' value chains. Green issues are being added as a new competition pattern.

Porter and Kramer (2011, 2006, 1999) enlarge its strategic framework by introducing a social dimension for competitiveness and using innovation as the essential mechanism for differentiating companies. Social and competitive patterns evolve from strategic philanthropy to shared value concepts through proactive corporate social responsibility. It is important to observe the movement from reactive behavior to a proactive and embedded one, from individual value chains to complex value chain networks, from shareholder governance to a rich set of multiple stakeholders. Companies are being asked anymore to be good corporate citizens but to be genuinely integrated into the complex structure of the socio-economic systems.

M.E. Porter grows its market-based analysis by reusing and evolving some basic frameworks and concepts related to the value chain and competitive advantage. Competitiveness is closed related to differentiation. Generic strategy and innovation is the core capability supporting this competitive process.

C.K. Prahalad's contribution is founded on a resource-based approach to sustainability. He raises his competitive analysis based on strategic concepts related to strategic intent and core competences (Hamel and Prahalad, 1989, 1994; Prahalad and Hamel, 1990; Prahalad, 1993). Prahalad and Ramaswamy (2004a, 2004b, 2003) evolve core competence concepts for co-creation and experience innovation perspectives, which enlarge the view of individual companies for cooperation and networking.

Sustainability appears in C. K. Prahalad's research when approaching 'poverty' issues. Hammond and Prahalad (2004) and Prahalad and Hart (2002) show companies that have been successful in making a profit while helping to contribute to the eradication of worldwide poverty by innovating not only in product-service concepts but in the whole set of operations networks.

Lubin and Esty (2010) describe sustainability as a megatrend as IT and quality were in the past few decades. For them, sustainability will touch every function, every business line, and employee. The lessons from IT and quality megatrends show that enterprises that intend to differentiate through sustainability would have to solve two concurrent problems: formulate a strategic vision for value creation and execute it. According to the authors, creating a strategic vision could be organized in four stages: 'do old things in new ways; 'do new things in new ways; 'transform a core business'; 'new business model creation and differentiation. Execution must be careful in 'elevating leadership,' 'systematizing methods and models,' 'aligning strategy and deployment,' 'integrating management,' and 'systematizing reporting and communication.' Vision creation is managed as a change process, and execution looks for coherent management and organizational structure. A sustainability performance system does vision and execution integration. The present rationality could be the lens for analyzing present 'business cases' for integrating sustainability into enterprise systems.

Sustainability has been placed on companies' strategic agenda. They are now looking for guidance to develop sustainability competences and integrate them into a complex, global, distributed, and dynamic operations network (Gunasekaran and Ngai 2012).

Sustainability in operations management was first approached through the triple bottom line model; that is, the economics, environmental, and social perspectives delimited operations systems studies (Veleva and Ellenbecker 2001). It was verified that sustainability is not an individual company issue but should be considered in the entire operations network (Kleindorfer et al. 2005). The next challenge that needs to be faced by sustainable operations models is sustainable initiatives coordination, and this creates an opportunity for governance discussion (O'Connor 2006; Seuring and Muller 2008). The present research on developing theoretical models for sustainable operations management focuses on integrative issues among triple bottom line perspectives and operations network actors (Bommel 2011; Dao et al. 2011; Talbot et al. 2011).

Besides sustainable operations models' content evolution, a set of model propositions that are oriented to establish a path for sustainable operations development could be selected. The Lowell Center for Sustainable Production, the University of Massachusetts at Lowell, MA, USA, supported some interesting studies for organizing sustainable operations evolution into five maturity levels that could be seen materialized in real-world applications (Veleva and Ellenbecker 2001; Veleva et al. 2001, 2003).

Following theory development that has received special attention from the academic community since the mid-80s, process frameworks are being proposed and applied by professional associations and consulting companies. Recently, it could be seen several models being developed and tested in real situations as the Industrial Research Institute (IRI) Sustainability Maturity Model that is being tested by 20 companies or consulting companies' maturity models as those proposed by Deloitte, FairRidge Group, SAP/ATOS, and AMR Research among others.

The main purpose of this chapter is to characterize these professional sustainability management models through the lens of the current sustainable operations theory. The research approach focuses on documents that describe the selected professional maturity models, developing a content analysis for identifying the main characteristics that define maturity levels and the mechanisms for evolution through the proposed levels (Flynn 1990; Montabon et al. 2007).

Drake and Spinler (2013) note that many decisions that determine the impact on a company's sustainability naturally intersect with flows of operations management, such as product design, technology choice, and supply chain management. It is part of the scope of sustainable operations management to analyze those decisions to identify opportunities for improvement and new processes.

Koehler and Hespenheide (2013) note that environmental, social, and governance (ESG) issues can affect the financial performance of companies, their operations or products directly or indirectly, through stakeholder actions along the value chain, in terms of risk of direct operations (accidents/equipment failures, carbon emissions, water pollution, employee strikes, health, and safety), supply chain risk (child labor, use of natural resources, climate disasters), product risk (toxic chemicals), genetically modified organisms, among others. According to the authors, to capture the actual value of ESG management, business leaders must demonstrate to their investors how they are being prepared to deal with ESG risks. This may suggest some changes in business strategy, including product innovation and business model redesign. They recognize that this could be a narrative of crisis and consequences. However, a closer look at ESG indicators would build a better, stricter future and make companies more valuable. The following figure illustrates the ESG indicators, and their integration implies social responsibility actions.

The collected data were used in comparative Tables, whose structure used definitions of CMM/CMMI and Pettigrew's (2012) framework for organizational analysis. CMMI structure is defined by generic and specific goals that work as design recommendations, sustainability practices, levels progression rules, and examples are studied in context, content/structure, and process perspectives.

10.2 Business Context for Sustainable Operations

Several survey results have been presented and discussed in the last years regarding companies' experience and perceptions in developing a sustainable business. It is important to observe that these findings somehow represent theory testing because they represent practical implementations that follow consolidated models. Evidence of changes in strategy, organizational structure, and adopted practices show how companies address their business through a sustainability lens.

In 2010, the World Economic Forum's Center for Global Growth Companies developed, in partnership with The Boston Consulting Group, the "Sustainability through Innovation" project. The study explained how emerging economies companies create innovative and profitable actions to drive growth while positively influencing local and global sustainability. The project showcases 16 companies named "New Sustainability Champions" that successfully turn constraints into opportunities through innovation, embed sustainability in their company culture, and actively shaped their business environments. Innovation is characterized by addressing a lack of resources, educating customers, and providing customers with appropriate financing. To embed sustainability could mean defining a bold sustainability vision, integrating sustainability into operations, and engaging the workforce in sustainability. To shape the business environment proactively influence policies and standards, partner to achieve mutual goals, and build awareness of sustainability importance (WEF 2011).

Koehler and Hespenheide (2013) observe that environmental, social, and governance (ESG) issues can impact companies' financial performance directly from their operations or products or indirectly through stakeholder actions along the entire value chain, in terms of direct operations risk (accidents /equipment failure, carbon emissions, water pollution, employee strikes, and health and safety), supply chain risk (child labor, natural resource use, weather catastrophes), product risk (toxic chemicals, genetically modified organisms recalls). According to these authors, to capture real value from ESG management, business leaders should demonstrate to their investors how they are being prepared to deal with ESG risks. This may suggest some changes in business strategy, including new product innovation and business model redesign. They acknowledge that this could be a narrative of crisis and aftermath; however, a closer look at ESG indicators would build better, more resilient, and more valuable enterprises. This Deloitte white paper discusses how important it is for the company to be prepared to manage sustainability issues. This could imply reviewing the entire conception of its business model.

KPMG International (2012) launched a sustainability report at its Global Summit in New York in February 2012. The information sets out a system of ten sustainability mega forces that will impact every business over the next 20 years: climate change, energy and fuel, material resource scarcity, population growth, wealth, urbanization, ecosystem decline, water scarcity, food security, and deforestation. It is observed that sustainable development and growth require action from both supplies and demand economic sides. The report proposes some necessary business actions: to understand

and assess risks; to use integrated strategic planning and strategy development; to turn strategic plans into ambitious targets and actions; to measure and report on sustainability; to seek collaboration with business partners on sustainability issues; to build strategic partnerships, seeking for genuine dialogue with governments and to demonstrate new and innovative approaches to Public–Private Partnerships. In this report, green issues represent most sustainability concerns and put some challenges for companies: developing energy and resource-efficient operations, sustainable supply chain management, strategic sector partnerships, and investing in sustainable product and services innovation.

Kiron et al. (2012) show some interesting results in their third-year consecutive survey on developing and implementing sustainable business practices. The research is a joint initiative of MIT Sloan Management Review and the Boston Consulting Group and, in 2011, surveyed around 4000 executives and managers from 113 countries. Results pointed out that 70% of companies have permanently placed sustainability in their strategic agendas, many of them in the last six years. Two-thirds of the respondents consider sustainability a necessary condition for being competitive in their served markets. Research analysis identified a group called 'Harvesters' (31% of total respondents) that are making a profit through their sustainability activities; in those cases, the changes in their operations frameworks and strategies are evident. Almost 25% of survey respondents could establish a connection between innovation and sustainability, covering product, process, and organizational aspects. The group of 'Harvesters' shows some interesting patterns regarding organizational structure (structures, lines of communications, performance metrics), business model (innovative and proactive approaches), and operations (integration and collaboration). Companies are definitively engaged in sustainability; they must combine senior leadership and integrated management to develop sustainable-based strategies, partner with organizations outside their businesses, and recognize that adopting sustainability practices takes time for learning and developing sustainable management competences.

GlobeScan and SustainAbility are running a research-based partnership intending to make a lasting contribution to sustainable development. They published a paper on the eve of the Rio+20 Earth Summit, offering an assessment of the present state of global opinion on sustainable development. They surveyed citizens (24,441 persons from 23 countries) and sustainability experts (1603 experts from 117 countries), presenting an accurate understanding of what people and experts think about sustainability leadership. Experts' survey shows that the best way for companies to contribute to sustainability are in order of importance related (Clinton 2012):

- Contributing to technological development and innovation.
- Working with governments to establish a regulatory environment that supports sustainable development.
- Improving internal sustainability performance.
- Influencing customers toward positive behavior change.
- Participating in multi-sectorial partnerships.
- Mobilizing suppliers on sustainability initiatives.

- Engaging employees on sustainability initiatives.

Using data from the same expert survey, Clinton and Whitaker (2012) identify barriers to business's transition to sustainability that are related in order of importance to:

- Financial short-termism.
- Inappropriate regulations.
- Low awareness of business imperative among business leaders.
- Low consumer demand for green business practices, products, and services.
- Lack of international standards.
- Lack of effective management tools/methodologies for business use.

According to Eccles et al. (2012), sustainability surveys show that many companies have come to consider sustainability-related strategies necessary to be competitive. Companies with an established organizational culture that includes strong capabilities for change, innovation commitment, and high levels of trust have a significant advantage in becoming sustainable companies. Eccles et al. (2012) also pointed out that employees in sustainable companies have a high level of confidence in each other, which implies taking the necessary risks to innovate and change their behaviors to support sustainability.

It could be said that the debate involves different actors such as government, organized civil society, not-for-profit organizations, companies, and academics. It is clear that at some points, consensus arises, influencing regulations, standards, reference models, design recommendations, etc. It could be pointed out:

1. Sustainability is a complex issue that covers the entire set of operations; it should be addressed from an operations chain or network perspective.
2. Risk management perspective.
3. Sustainable product and services innovation
4. Vision (strategic component, proactive) and execution (capabilities) integration. Process-based approach and PMS (assure deployment, coordination, and integration).
5. Sustainability culture.

10.3 Sustainable Operations Models

OM research in sustainability is a growing issue, and it is generating new theories and models for OM practice. First, sustainability could not be achieved by a single firm's action. Research has shown that for sustainability to be truly effective, entire supply chains must operate sustainably, not just individual partners (Carter and Rogers 2008; Kleindorfer et al. 2005).

Ueda et al. (2009) highlight the importance of an appropriate approach to sustainability issues in designing, implementing, and running enterprise systems. They ask for an integrative approach and new value definitions for a society conscious of its role. Models should be developed and tested at the level of operations function and the entire value chain system. Another critical point is approaching sustainability issues through the lens of multiple stakeholders. Multiple views could lead to different structures and processes for sustainability integration.

The 'Lowell Center for Sustainable Production—LCSP' defines nine principles of sustainable production (Veleva and Ellenbecker 2001):

- Products and packaging are designed to be safe and ecologically sound throughout their life cycles; services are designed to be safe and environmentally sound.
- Wastes and ecologically incompatible by-products are continuously reduced, eliminated, or recycled.
- Energy and materials are conserved, and the forms of energy and materials used are most appropriate for the desired ends.
- Chemical substances, physical agents, technologies, and work practices that present hazards to human health or the environment are continuously reduced or eliminated.
- Workplaces are designed to minimize or eliminate physical, chemical, biological, and ergonomic hazards.
- Management is committed to an open, participatory continuous evaluation and improvement process focused on the firm's long-term economic performance.
- Work is organized to conserve and enhance the efficiency and creativity of employees.
- The security and well-being of all employees is a priority, as is the continuous development of their talents and capacities.
- The communities around workplaces are respected and enhanced economically, socially, culturally, and physically; equity and fairness are promoted.

LCSP defines sustainable production as the creation of goods and services processes and systems that are: non-polluting; conserving energy and natural resources; economically viable; safe and healthful for workers, communities, and consumers; and socially and creatively rewarding for all working people (Veleva et al. 2001).

Figure 10.1 synthesizes OM sustainability models' evolution over the last ten years, illustrating this process with some not exhaustive studies. It is suggested for characterizing models' evolution, from 3BL applications, through Extended 3BL and 3BL plus governance to reach the integrated 3BL approach. Finally, Eriksson and Svensson (2015) understand that the organization has internal drivers to develop its social responsibility: a comprehensive and holistic approach, management support, functional and strategic responsibility, sustainability incentives, measurement and reporting system, and corporate education. Social responsibility and internal silos can face integration and coordination problems. The role of supply chain management is to coordinate all social responsibility efforts between companies that create social and environmental strategies. There are several strategies, such as the need for purchasing personnel and quality of supply to visit supplier factories, to integrate the purchasing department's social responsibilities policies. Companies need to take a shared and inter-functional approach to social responsibility. The holistic inner

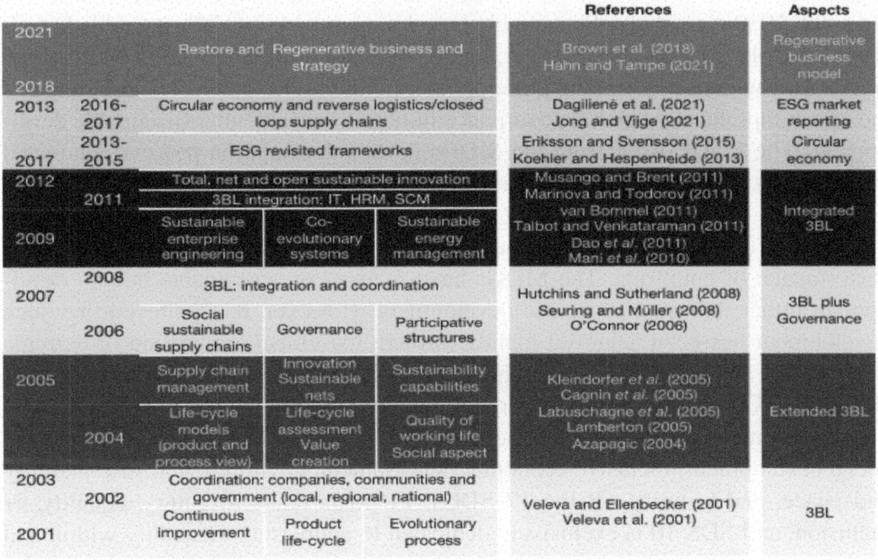

Fig. 10.1 OM sustainability management framework evolution

vision element both drives and facilitates social responsibility efforts. Companies that want to implement social responsibility need to allocate resources to develop and ensure that the purchasing team has the appropriate capabilities to evaluate and work with suppliers.

Eriksson and Svensson (2015) also point out that managers need to prioritize social responsibility, as their actions drive and facilitate social responsibility efforts. Functional or strategic responsibility is contingent on management support and affects individuals in a specific way. Awareness of business sustainability expectations should be created and purchasing managers should be considered in decision-making. However, it should not be just responsible for making decisions without reward. Employees need both support and a mandate to opt for socially accountable choices, even when they are expensive. If business sustainability is on the strategic plan, incentives must link employee behaviors to growing and improved efforts in business sustainability. The "measurement" of social responsibility efforts at the organizational level is also advocated. Social responsibility goals must be linked to organizational goals. Social responsibility efforts should be part of the leading measurement indicators. What is not measured is not tangible. Measuring the element drives social responsibility. Corporate education includes the need for employees to be trained in social responsibility. Training issues include environment and working conditions. Employees need to understand what they do and the results of their actions.

According to Jong and Vijge (2021), since the publication of the Brundtland report of 1987, the socio-economic paradigm of sustainable development has been highly influential in global and national frameworks. Among the various initiatives that stand out is the 2030 Agenda for Sustainable Development with its 17 Sustainable Development Goals (SDGs). The 2030 Agenda is an evolution of the 2000 Millennium Declaration with its Millennium Development Goals (MDGs), which is recognized as the first comprehensive global governance initiative for achieving sustainable development. The SDDs differ from the MDDs in their formulation process conducted through multi-stakeholder consultations around the world.

Jong and Vijge (2021) note that in the decades before the introduction of the MDGs, the international development agenda focused mainly on the ecological performance of countries. The MDGs have focused on sustainable human development rather than just economic development. However, it is believed that such development has been achieved mainly through the global eradication of extreme poverty. The authors highlight that the SDGs differ in several key elements and, that is, the SDGs have broadened the predominantly singular objective of the MDGs: the reduction of extreme poverty; to include four objective dimensions: inclusive social development, inclusive economic development, environmental sustainability and peace, and security. Of the 17 SDCs, 11 goals relate to equity, equality, or inclusion, and SDS 10 is exclusively dedicated to addressing inequality within and between countries.

Jong and Vijge (2021) also highlight that the SDGs present a much more integrated approach to sustainable development than the MDGs, focused on the interaction between the 5 Ps of people, prosperity, planet, partnership, and peace. The SDGs are created as "indivisible and interconnected," with a strong focus on "political and institutional coherence" and coordination in all sectors and policies to achieve sustainable development's three economic, social, and environmental dimensions. Instead of being subordinated to the economic dimension, sustainable development's social and ecological dimensions are at the heart of the SDGs.

Dagilienė et al. (2021) point out that the combined use of technology and sustainable operations practices in production and consumption models has contributed to creating a circular economy model. A circular economy is often presented as a practical way to implement sustainable development. The circular economy puts the environmental pillar of sustainability in the spotlight and recognizes the need for favorable economic environments, but social sustainability is seen as consequential. The circular economy may be the expression of a soft approach to sustainability. It suggests a system where recycling and reuse are assumed as substitutes for raw materials. The urgency behind a circular transformation is based on an impending resource crisis and globalization, and increased consumption of these resources. In practice, the circular economy is implemented in various contexts, from resource efficiency, product design, management and proper recycling of waste, practices of companies for the reuse and repair of raw materials and equipment, transformation of post-consumer waste into valuable materials urban circular initiatives.

According to Brown et al. (2018), sustainability should not be seen as a journey, but a state of equilibrium, An essential element of moving from sustainability to regenerative is education, and it must be seen as the basics of all sustainability strategies, that entails shifting from the individual point of view and perception toward a collaborative, cooperative, and responsible approach. To operationalize the notion of regenerative business, Hahn and Tampe (2021) develop a range of regenerative strategies: restore, preserve, and enhance.

Created the evolutionary context regarding OM models and sustainability, an analysis could be developed for professional or practical sustainability models that companies are adopting.

10.4 Professional Maturity Models

Leading professionals are also involved in creating solutions for enterprises' sustainability management. Consultancy firms and professional associations are proposing some sustainability maturity models. These models are founded on a process view and focus on implementation issues.

Industrial Research Institute (IRI) is developing a maturity model for assessing and guiding companies in creating innovative sustainable products and services that drive growth and contribute to the triple bottom line. The model is structured in four primary stages: beginning, improving, succeeding, and leading. New product development and DfE Design for environment found the developed model. Table 10.1 shows some characteristics regarding operations management (Brandt et al., 2012).

Deloitte Consulting is developing a "Maturity model for sustainability efforts" model. The model establishes four stages for sustainability maturity development: maturity level 1—reactive or little response; maturity level 2—managing direct risks and costs; maturity level 3—established strategy and stakeholder collaboration; maturity level 4—integrated strategy for competitive advantage. It is assumed that sustainability is a strategic business issue, and, to be successful, companies must consider incorporating it into their core business models. Companies approaching sustainability through ad hoc programs or isolated initiatives do not achieve their desired goals as effectively as companies with an integrated approach (Park and Pavlovsky 2010).

Fairbridge Group is a team of management, strategy, and change experts focused on business transformation through the practical application of sustainability for operational improvement and strategic innovation. They developed a framework for sustainability management that integrates strategy, operations, branding, measurement, and organizational development. The framework is called the "Sustainability Management Maturity Model (SM3)", which allows organizations to assess how ready their infrastructure is to address sustainability challenges. The model has five levels of maturity (Scott 2009):

- Business as Usual: The company has some awareness of sustainability concerns but is not doing anything to address them.
- Ad Hoc Response: The company is starting to respond to questionnaires from customers and NGOs but does not have a formal sustainability function.
- Plan and Pilot: The company has created a sustainability department that has started planning and piloting some eco-efficiency projects.

Table 10.1 IRI maturity model—OM issues

Category	Beginning	Improving	Succeeding	Leading
Design for Environment (DfE) **Supply Chain**	• Product design teams use the supply chain to procure what is specified • Suppliers are capable of supplying material meeting specific regulations and directives	• Suppliers are involved early in product design activities related to sustainability and environmental responsibility	• Supplier involvement in design activities results in using "Greener" materials and processes • Environmental impact of transportation is used in selecting suppliers	• Suppliers are actively engaged in the Ecodesign process; they understand how they affect the finished product's environmental footprint and make improvements
Design for Environmental (DfE) **Manufacture**	The organization is beginning to assess the environmental impact of manufacturing • Designers begin to understand how design choices affect the environmental impact of manufacturing	Manufacturing impact becomes a criterion in product design • The impact of product design on manufacturing choices is understood • Product designers begin to design new products to reduce manufacturing impact	Reduced environmental impact of manufacturing is a mandatory design criterion in the NPD process •LCA results are used to assess the overall environmental impact • Product design is optimized to reduce the environmental impact •Transportation is included in LCA	The organization is an industry leader in innovation in design for manufacturing to reduce life cycle environmental impact • Works with manufacturing equipment suppliers to reduce the environmental impact of manufacturing

Source Brandt et al. (2012)

- Operationalize: The company's lines of business have taken ownership of sustainability initiatives and have started to leverage sustainability to extend their products and services portfolio.
- Transform: The company has a system-wide sustainability mandate fully integrated into all aspects of the business; the resulting transformation creates clear competitive differentiation.

ATOS and SAP consulting had an agreement to develop a sustainability maturity assessment model. The model is organized in five stages: maturity level 1—compliance; maturity level 2—corporate social responsibility; maturity level 3—process change; maturity level 4—continuous improvement; maturity level 5—environmental excellence. Incorporating sustainability into business practices is a significant challenge for many organizations. Customers, government, and other stakeholders exert pressure on businesses to meet ever more stringent targets. All expect to see the real, demonstrable value and not 'greenwash.' This is how SAP approaches sustainability through continuous competences development (ATOS 2011).

AMR Research frame its Sustainability Maturity Model into five stages: sustainability not in focus (compliance), tactical sustainability (reporting), strategic transformation, structural normalization, and operational implementation. The last three are named strategic sustainability and define a proactive and innovative approach for designing and managing sustainable operations. The following comment synthesizes AMR consciousness regarding sustainability (Stokes 2008):

> Despite the logic of fully and successfully embracing sustainable business practices, the precise definition of exactly what constitutes best-of-breed sustainable performance remains unclear. Previous models, such as triple bottom line, provide sound, aspirational philosophies, but a holistic roadmap or framework that fully embraces sustainable organizational transformation has not been successfully communicated. Without clarity, the emergence of true leadership and attendant benefits are correspondingly enigmatic.

Some concepts regarding maturity could be borrowed from CMMI, which has an extensive experience in studying such phenomena in software development industry. According to SEI (2010):

> A maturity level consists of related specific and generic practices for a predefined set of process areas that improve the organization's overall performance. The maturity level of an organization provides a way to characterize its performance. Experience has shown that organizations do their best when they focus their process improvement efforts on a manageable number of process areas and that those areas require increasing sophistication as the organization improves. A maturity level is a defined evolutionary plateau for organizational process improvement. Each maturity level matures an important subset of the organization's processes, preparing it to move to the next maturity level. The maturity levels are measured by achieving the specific and generic goals associated with each predefined set of process areas.

It is also important to set mechanisms for maturity levels transition, and according to SEI (2010) experience:

> Organizations can achieve progressive improvements in their maturity by achieving control first at the project level and continuing to the most advanced level - organization-wide performance management and continuous process improvement - using qualitative and quantitative data to make decisions. Since improved organizational maturity is associated with improvement in the range of expected results that an organization can achieve, maturity is one way of predicting general outcomes of the organization's next project.

Understood maturity levels transition mechanisms concepts, it is essential to delimit sustainability models analysis, and Pettigrew's (2012) organizational analysis framework could be used for that purpose. His analysis is founded in three primary perspectives: context, content/structure, and processes.

Pettigrew's (2012) approach emphasizes is on describing and analyzing processes of change in context. Context power as action and its outcomes explanation is enabled by the treatment of context as an 'interactionist' field of multiple analysis levels. Pettigrew (1987) refers to the outer context as the social, economic, political, and competitive environment in which companies operate. Inner context refers to the firm's structure, corporate culture, and political context through which ideas for change have to proceed.

Content of any strategic change is ultimately a legitimization process result, shaped by changes in the organization's outer context and by political and cultural considerations inside them, though often expressed in analytical terms. This rationality still has a significant impact on practical strategies for intervening to create change in organizational settings. Simple linear formulating and implementing strategy notions must contend with newly exposed complexity levels. Content refers to transformation understudy of areas of interest. A company could be seeking to change technology, organizational design, products, services, production processes, or corporate culture (Pettigrew 2012, 1987).

A process approach for organizational analysis must go beyond identifying and explaining processes patterns and demonstrate how and to what extent processes can and do shape outcomes. The method of change refers to actions, reactions, and interactions from various interested actors as they seek to move companies from their present positions to a desirable future state (Pettigrew 2012, 1987).

Table 10.2 compares the selected professional, sustainable maturity frameworks regarding sustainability maturity concepts, CMMI framework adherence, and Pettigrew's scope definition. The Table is based on 'textual' evidence found in model documentation provided by public sources. This research aims to find the existence of such evidence but not to assess its content.

Although theoretical concepts and models support professional initiatives, implementation is sometimes incomplete. It is essential to observe that the entire set of practical frameworks primarily focuses on environmental aspects, but strategic integration and approach, continuous improvement orientation, risk analysis, innovation-based evolution, and multiple stakeholders are great content. The studied frameworks do not directly approach economic sustainability.

As a synthesis, it could be said:

- Sustainability is a systemic phenomenon and should be approached by the entire relationships in operations systems.
- For operations strategy models and their system of aspects, an economic perspective is a mature issue, the environment is being consolidated, but social factors are being developed.
- It is required a revision of value definition since firm definition through the entire network of relationships.
- Implementing and integrating a sustainable view for operations strategy could be done through a life cycle and a maturity-based model.

Table 10.2 Sustainable maturity models content analysis

Security maturity model	Sustainability maturity concepts		CMMI adherence					Scope definition		
	Maturity levels (Reference = 5)	TBL	Generic goals	Specific goals	Practices	Examples	Maturity levels transition	Content	Context	Process
IRI	++	+++	+++	—	—	—	—	+++	—	—
Deloitte	++	+++	—	—	—	—	+++	—	—	—
Fair Ridge Group	+++	—	++	—	—	++	+++	++	++	++
ATOS and SAP	+++	—	—	—	—	—	+	—	—	—
AMR Research	++	+++	+++	—	—	—	+++	+++	+++	—

It could be seen on professional frameworks content analysis that sustainability fundamentals of an emergent theory for sustainable operations are clearly defined and specified for practical application, however, scoped delimitation and governance structures are not formally addressed.

Maturity levels tend to be organized in five groups that forms a continuous development path from 'compliance' positioning to 'strategic' operations network integration. Capabilities or competences could describe the 'operations vision' associated with each maturity level.

COVID 19, the acceleration of digital transformation processes, join a growing demand for sustainable economic development. One cannot talk about sustainable development without mentioning its 17 objectives. The Sustainable Development Goals (SDGs) were established by the United Nations (UN) in 2015 and form a global agenda for the construction and implementation of public policies aimed at guiding humanity until 2030. The ODS Brazil (2021) explains that the schedule includes an international action plan to achieve the 17 SDGs, unfolded in 169 goals, which address several fundamental themes for human development in five perspectives: people, planet, prosperity, partnership, and peace.

The main contribution of this chapter is to construct a better understanding of sustainability maturity-based models management through the analysis of process frameworks that sustainable companies are using.

Acknowledgements The authors wish to thank CNPq (National Council of Technological and Scientific Development) for grant 317150/2021-9.

References

ATOS (2011) Sustainability maturity assessment model. Factsheet, June-July, London, UK

Brandt V, Hynds J, Taylor J, Burek S, Yager W, Penton H, Ward S, Schwartz L (2012) Sustainability maturity model. Industrial Research Institute, 26-28 May, Arlington, USA

Brown M, Haselsteiner E, Apró D, Kopeva D, Luca E, Pulkkinen K, Vula Rizvanolli B (eds) (2018) Sustainability, restorative to regenerative. EURAC Research, Institute for Renewable Energy, Bolzano, Italy

Carter CR, Rogers DS (2008) A framework of sustainable supply chain management: moving toward new theory. Int J Phys Distrib Logist Manag 38(5):360–387

Clinton L (2012) Down to business: leading at Rio+20 and beyond. SustainAbility, London

Clinton L, Whitaker K (2012) Unfinished business: perspectives from the sustainable development frontier. SustainAbility, London

Dagilienė L, Varaniūtė V, Bruneckienė J (2021) Local governments' perspective on implementing the circular economy: a framework for future solutions. J Clean Prod 310:127340

Dao V, Langella I, Carbo J (2011) From green to sustainability: information technology and an integrated sustainability framework. J Strateg Inf Syst 20(1):63–79

Drake D, Spinler S (2013) Sustainable operations management: an enduring stream or a passing fancy? Manuf Serv Oper Manag 15(4):689–700

Eccles RG, Perkins KM, Serafeim G (2012) How to become a sustainable company. MIT Sloan Manag Rev 53(4):42–50

Eriksson D, Svensson G (2015) Elements affecting social responsibility in supply chains. Supply Chain Manag 20(5):561–566
Flynn B (1990) Empirical research methods in operations management. J Oper Manag 9(2):250–284
Gunasekaran A, Ngai EWT (2012) The future of operations management: an outlook and analysis. Int J Prod Econ 135(2):687–701
Hamel G, Prahalad CK (1989) Strategic intent. Harv Bus Rev 67(3):63–76
Hamel G, Prahalad CK (1994) Competing for the future. Harv Bus Rev 72(4):122–128
Hammond A, Prahalad CK (2004) Selling to the poor. Foreign Policy 142:30–37
Hahn T, Tampe M (2021) Strategies for regenerative business. Strateg Organ 19(3):456–477
Jong E, Vijge MJ (2021) From millennium to sustainable development goals: evolving discourses and their reflection in policy coherence for development. Earth Syst Gov 7:100087
Kiron D, Kruschwitz N, Haanaes K, von Streng Velken I (2012) Sustainability nears a tipping point. MIT Sloan Manag Rev 53(2):69–74
Kleindorfer PR, Singhal K, van Wassenhove LN (2005) Sustainable operations management. Prod Oper Manag 14(4):482–492
Koehler DA, Hespenheide EJ (2013) Finding the value in environmental, social, and governance performance. Deloitte Rev 12:97–111
KPMG International (2012) Expect the unexpected: building business value in a changing world. Sustain Insight 120364:1–18
Lubin DA, Esty DC (2010) The sustainability imperative. Harv Bus Rev 88(5):42–50
Montabon F, Sroufe R, Narasimhan R (2007) An examination of corporate reporting, environmental management practices and firm performance. J Oper Manag 25(5):998–1014
O'Connor M (2006) The "Four Spheres" framework for sustainability. Ecol Complex 3(4):285–292
Park C, Pavlovsky K (2010) Sustainability in business today: a cross-industry view. Deloitte Sustainability White Papers, New York, USA
Pettigrew AM (1987) Context and action in the transformation of the firm. J Manage Stud 24(6):649–670
Pettigrew AM (2012) Context and action in the transformation of the firm: a reprise. J Manage Stud 49(7):1304–1328
Porter ME (1979) How competitive forces shape strategy. Harv Bus Rev 57(2):137–145
Porter ME (1987) From competitive advantage to corporate strategy. Harv Bus Rev 65(3):43–59
Porter ME (1990) The competitive advantage of nations. Harv Bus Rev 68(2):71–91
Porter ME (1991) America's green strategy. Sci Am 264(4):168
Porter ME, Esty D (1998) Industrial ecology and competitiveness: strategic implications for the firm. J Ind Ecol 2(1):35–43
Porter ME, Kramer MR (1999) Philanthropy's new agenda: creating value. Harv Bus Rev 77(6):5–16
Porter ME, Kramer MR (2006) Strategy and society: the link between competitive advantage and corporate social responsibility. Harv Bus Rev 84(12):1–14
Porter ME, Kramer MR (2011) Creating shared value: how to reinvent capitalism and unleash a wave of innovation and growth. Harv Bus Rev 89(1–2):1–17
Porter ME, van der Linde C (1995a) Green and competitive: ending the stalemate. Harv Bus Rev 73(5):120–134
Porter ME, van der Linde C (1995b) Toward a new conception of the environment-competitiveness relationship. J Econ Perspect 9(4):97–118
Prahalad CK, Ramaswamy V (2004a) Co-creating unique value with customers. Strat Leadersh 32(3):4–9
Prahalad CK (1993) The role of core competencies in the corporation. Res Technol Manag 36(6):40–47
Prahalad CK, Hart S (2002) The fortune at the bottom of the pyramid. Strat Bus 26:54–67
Prahalad CK, Hamel G (1990) The core competence of the corporation. Harv Bus Rev 68(3):79–87
Prahalad CK, Ramaswamy V (2003) The new frontier of experience innovation. MIT Sloan Manag Rev 44(4):12–18

Prahalad CK, Ramaswamy V (2004b) Co-creation experiences: the next practice in value creation. J Interact Mark 18(3):5–14
Scott J (2009) Sustainability management infrastructure: what it is and why you should care. Fair-Ridge Group, Provo, USA. Availabe at http://www.triplepundit.com/2009/06/sustainability-management-infrastructure-what-it-is-and-why-you-should-care. Accessed on 5 Apr 2013
SEI (2010) Capability maturity model integration—CMMI® version 1.3. Software Engineering Institute—CMMI® Product Development Team, Pittsburgh, USA
Seuring S, Müller M (2008) From a literature review to a conceptual framework for sustainable supply chain management. J Clean Prod 16(15):1699–1710
Stokes S (2008) Crossing the great divide: Sustainability as corporate strategy, Sustainability strategies report. AMR Research Inc., Boston, USA
Talbot J, Venkataraman R (2011) Integration of sustainability principles into project baselines using a comprehensive indicator set. Int Bus Econ Res J 10(9):29–40
Ueda K, Takenaka T, Váncza J, Monostori L (2009) Value creation and decision-making in sustainable society. CIRP Ann Manuf Technol 58(2):681–700
van Bommel HWM (2011) A conceptual framework for analyzing sustainability strategies in industrial supply networks from an innovation perspective. J Clean Prod 19(8):895–904
Veleva V, Ellenbecker M (2001) Indicators of sustainable production: framework and methodology. J Clean Prod 9(6):519–549
Veleva V, Hart M, Greiner T, Crumbley C (2001) Indicators of sustainable production. J Clean Prod 9(5):447–452
Veleva V, Hart M, Greiner T, Crumbley C (2003) Indicators for measuring environmental sustainability: a case study of the pharmaceutical industry. Benchmark Int J 10(2):107–119
WEF (2011) Redefining the future of growth: the new sustainability champions. World Economic Forum, Geneva

Edson Pinheiro De Lima Received his Ph.D. in Industrial Engineering from Federal University of Santa Catarina (UFSC), his master's degree in electrical engineering (Automation) from University of Campinas (UNICAMP) and his Bachelor of Science in Electrical Engineering from Federal University of Technology—Parana (UTFPR). He is a Professor at the Industrial and Systems Engineering of Federal University of Technology—Parana (UTFPR), Brazil. During his Ph.D., he spent 12 months as a Visiting Researcher at the Logistics and Production Engineering Laboratory of the Production Engineering, Business and Statistics Department of the Industrial Engineering School at Polytechnic University of Madrid, Spain. Professor Edson Pinheiro is being over the years a Visiting Scholar in the Warwick Business School (2007), Polytechnic University of Madri (2010) and ISE at Virginia Tech (2017–2018). His current research interests include sustainable operations, operations strategy, performance measurement systems and performance analytics.

Sergio E. Gouvea Da Costa Received his Ph.D. in Industrial Engineering from University of São Paulo (USP), his master's degree in electrical engineering (Automation) from University of Campinas (UNICAMP) and his Bachelor of Science in Electrical Engineering from Federal University of Technology—Parana (UTFPR). During his doctorate, he carried out research at the Institute for Manufacturing da University of Cambridge, England (2000–2001). He did Postdoctoral work at the Edward P. Fitts Department of Industrial and Systems Engineering at North Carolina State University, USA (2009–2010). He is an Associate Professor at the Industrial and Systems Engineering of Federal University of Technology—Parana (UTFPR), Brazil. His current research interests include Energy Management, Operations Strategy and MCDM/A.

Carla Gonçalves Machado Received his Ph.D. in Industrial and Systems Engineering from Pontifical Catholic University of Parana (PUCPR), his master's degree in industrial engineering from State University of Sao Paulo (UNESP) and his Bachelor of Science in Business/Corporate Communication—Public Relations from State University of Sao Paulo (UNESP). Between 2018 and 2020, Carla was Postdoctoral Researcher and Researcher at the Division of Supply & Operations Management involved in research projects on Sustainability, Digitalization and Supply & Operations Management, also teaching in Master courses—Operations Strategy; Manufacturing Strategies; Business Marketing and Purchasing. She is an Assistant Professor in Business Administration at Jönköping International Business School, Jönköping University, Sweden. She is Director of Societal Engagement at Jönköping International Business School. Her current research interests include sustainable operations management, digitalization, and supply chain management.

Pablo Carpejani Ph.D. candidate in Industrial and Systems Engineering at Pontifical Catholic University of Parana (PUCPR), his master's degree in industrial and systems engineering from Pontifical Catholic University of Parana (PUCPR), and his Bachelor of Science in Business Administration from Pontifical Catholic University of Parana (PUCPR). He is an Assistant Professor at the Business of Pontifical Catholic University of Parana (PUCPR), Brazil. He participated in sustainability reporting processes, from measuring performance, creating the materiality process, and publishing the report. He is currently coordinator of the MBA in corporate sustainability at PUCPR. His current research interests include sustainable operations, GRI reports, ESG models, and sustainability materiality.

Chapter 11
Sustainability Indicators Design Process for ESG

Edson Pinheiro De Lima, Pablo Carpejani, Carla Gonçalves Machado, Rafael A. Kluska, and Sergio E. Gouvea Da Costa

Abstract Sustainable development has defined the competitive context in the last decade, forcing companies to review their business model to incorporate sustainability requirements. Stakeholders demand more transparency reflected in the present regulatory framework formed by standards, guidelines, and performance reports. These elements provide information for markets and society concerning sustainable behavior, which companies have certified or formally agreed upon. Organizations' operations performance information could be related to evolutionary performance indicators set, representing sustainability present performance and orient companies to reach higher sustainability performance levels, following a maturity trajectory. This chapter aims to develop a process for designing an integrated set of performance indicators for managing sustainability at the operations system level. Applied research is conducted based on the 'Cambridge Process Approach' and worked in a research BPM cycle. Sustainability performance indicators are formally defined and detailed as metadata. Process application allows a better understanding of how sustainability indicators could be integrated and related to an ESG-based model and create an information system that could be used for auditing, assessment, and reporting sustainability performance.

E. P. De Lima (✉) · S. E. G. Da Costa
Industrial and Systems Engineering of Federal University of Technology, Curitiba, Parana (UTFPR), Brazil
e-mail: pinheiro@utfpr.edu.br

S. E. G. Da Costa
e-mail: gouvea@utfpr.edu.br

P. Carpejani
Business of Pontifical Catholic, University of Parana (PUCPR), Curitiba, Brazil
e-mail: p.carpejani@pucpr.br

C. G. Machado
Business Administration at Jönköping International Business School, Jönköping University, Jonkoping, Sweden
e-mail: carla.goncalvesmachado@ju.se

R. A. Kluska
Fluxo Business Automation, Curitiba, Brazil

M. J. Meiriño and O. L. G. Quelhas (eds.), *Critical Factors Associated with Environmental, Social and Governance Management in Organizations*, Accounting, Finance, Sustainability, Governance & Fraud: Theory and Application, https://doi.org/10.1007/978-981-96-1725-8_11

Keywords Sustainability · Sustainability indicators · ESG designing

11.1 Regulatory Framework

Business stakeholders are demanding more transparency related to companies' results and performance, particularly in those aspects connected to sustainability. Some evidence points in that direction as companies' shareholders and stockholders look for evidence that supports governance best practices, customers and clients are demanding information related to production raw materials and components origins and suppliers' labor conditions, and employees are looking for opportunities in companies that are socially responsible and aware of their environmental duties, governments and the society, in general, are creating 'global' requirements for companies to report their economic, environmental, and social performance (Reuter et al. 2012; Gold et al. 2010; Aras and Crowther 2008; Castka and Balzarova 2007; Keeble et al. 2003).

The last three decades cover a significant evolution of the regulatory framework for 'sustainability,' encompassing economic, environmental, and social aspects. Several documents are being used to create a global socio-economic system based on sustainable development principles.

Sustainability standards also experimented with an interesting evolution in the last decades and are responsible for guiding implementation in the sustainability regulatory framework. Some standards could be cited as representative of sustainability implementation: ISO 14000—environment (1993); ISO 14040—life cycle assessment (2001); SA 8000—social accountability (1997); OHSAS 18001—health and safety (1999); AA 1000—accountability (1999); ISO 26000—corporate social responsibility (2010); ISO 50001—energy management (2011) (Hundzinski et al. 2013; Castka and Balzarova 2007).

Accountability and reporting are essential activities for companies showing their sustainability initiatives. A sustainability report is a report published by an organization about the economic, environmental, and social impacts caused by its activities. Global Report Initiative (GRI), created in 1997, proposes a framework, that is, a reporting system, which provides metrics and methods for measuring and reporting sustainability-related impacts and performance. Its version 4.0, GRI, helps companies identify material sustainability issues (Hundzinski et al. 2013; Tate et al. 2010; Labuschagne et al. 2005).

Although reporting is a crucial activity for sustainability accounting, sustainability performance measurement develops the management component of sustainability. It was shown that the sustainability regulatory framework is complex and follows an evolutionary path, so companies should comply with these requisites to meet local and global demands. Hart and Milstein (2003) highlight that sustainability metrics and indicators should be approached as 'instruments' used for value creation; benefits should be clearly defined when adopting a sustainability-based business model.

Operations Management literature is full of sustainability metrics, measures/indicators, and measurement systems studies. Taking some examples, an evolutionary trajectory could be traced to how sustainability is assessed and related to companies' operations performance. There is a representative paper set in Meadows (1998); Azapagic and Perdan (2000); Veleva and Ellenbecker (2001); Keeble et al. (2003); Labuschagne et al. (2005); Hutchins and Sutherland (2008); van Bommel (2011); Golini et al. (2014), and Schrettle et al. (2014).

Companies are putting great effort into developing integrated systems for managing sustainability, coordinating all aspects related to sustainability in terms of economics, environmental, and social issues. Another point is that companies are approaching their operations considering their operations network covering multiple supply chains. In addition, sustainability capabilities are organized in maturity-based models, and companies for developing their systems over time (Maas and Reniers 2014; Looy et al. 2013; Cabezas et al. 2003).

Consultancy and software companies support the companies' effort to develop operations sustainability management systems. Several products are being offered: Enablon® and Credit 360® are designed to address the integrated management system; ToFuture® is GRI-based; and traditional ERP software companies such as SAP and Oracle.

11.2 Conceptual Framework Development

Operations management research on sustainability is strongly influenced by a context defined by sustainable development. The Brundtland report, written in 1987 by the World Commission on Environment and Development (WCED), defined sustainable development as "development that meets the needs of the present without compromising the ability of future generations to meet their own needs" (WCED 1987).

Isaksson and Steimle (2009) highlight that sustainable development is not only a concern for nations but also companies. They point out the importance of big corporations in driving sustainable development. The challenge presented for companies is not only defined by reducing pollution, but corporate social responsibility (CSR) represents enterprises' commitment to behave socially and environmentally responsible while achieving their economic goals. Hasna (2010) commented that companies should approach sustainability considering its multiple aspects and dimensions in a coordinated and integrated way; he also observed that the economic system is defined by sustainable development in value creation.

Some international surveys reveal that companies are integrating sustainability into their strategy, and some are being called 'harvesters' because their approach to sustainability results in performance. Governance and leadership constructed based on sustainability principles change how companies produce value for their stakeholders (Parisi 2013; Kiron et al. 2012).

Mori and Christodoulou (2012) show that the 'Triple Bottom Line (TBL) framework is essential for creating a balanced and integrated vision regarding sustainability. Van Bommel (2011); Ueda et al. (2009); Hutchins and Sutherland (2008); Porter and Kramer (2006), and Wilkinson et al. (2001) integrated sustainability into companies' strategy, value chain, and organizational design using TBL framework and also point out the necessity of coordination.

Labuschagne et al. (2005) show how difficult it is to express the concept of sustainability in concrete terms. The synergy proposed by the systems of available indicators still does not adequately meet the needs of decision-makers for operations analysis and assessment in terms of internal and external requirements.

Transforming sustainability strategies into actions requires new performance indicators to measure company performance, and the management model also needs to be redesigned. The design, structure, and management system must be aligned to coordinate actions and motivate the teams to implement sustainability (McCartney 2009; Epstein and Roy 2001).

According to Veleva et al. (2001), sustainability indicators are important in assessing companies' contribution to sustainable development, founded on their facilities and operations.

Epstein and Roy's (2001) framework show that each company is subject to different contexts and demands from multiple stakeholders. An appropriate set of indicators should be developed for managing these multiples into rests and relationships. According to Keeble et al. (2003), sustainability indicators should be continuously revised, supplying relevant information to decision-makers. Krajnc and Glavič (2005) suggest that sustainability performance indicators are grouped in a single platform to support decision-making.

To generate sustainability performance indicators, some criteria should be observed (Veleva and Ellenbecker 2001): (i) the indicators should be aligned with sustainable production strategy; (ii) be based on available and liable information; (iii) they should be verifiable; (iv) should be composed of key and complementary indicators; (v) contain a significant number of indicators; (vi) be easily applied; (vii) be made up of qualitative and quantitative indicators; (viii) should allow for comparisons; (ix) be close to the leading global issues; (x) be aligned with national and local sustainability indicators; (xi) be developed and evaluated with the participation of stakeholders.

The indicators could be based on validated and recognized sources as standards and other normative documents such as Global Reporting Initiative (GRI), ISO 9001, ISO 14001, OHSAS 18001, ISO 26000, ISO 14040, Agenda 21, Global Compact (Louette, 2008).

Thus, sustainability indicators must be logical and traceable to be replicated and comparable to assess sustainable operations performance. Furthermore, data collecting instruments should be certified to guarantee robustness and reliability; besides, sources of information and margins of error in measurement should be clearly expressed (Kleindorfer et al. 2005; Kennerley and Neely 2003). For this, sustainability indicators must be integrated horizontally throughout the company value chain and represent strategy deployment as a reflection of strategic positioning and objectives (Hart and Milstein 2003; Krajnc and Glavič 2005).

Value chain, a system of value chains, supply chains, and production networks, are organizational structures representing material and informational flow. Sustainability should be constructed through value creation and embedded in products and production processes. Sustainability indicators should be correlated to activities developed in companies' value chain (Leppelt et al. 2013; Ueda et al. 2009; Nidumolu et al. 2009; Porter and Kramer 2006).

According to Veleva et al. (2001), sustainable systems follow a continuous and evolutionary "[...] starting at different places in the evolutionary process". This means that a company can adapt its strategies and operations model to an evolving path toward high levels of sustainability. The companies should face the challenge of developing assessment and auditing systems that make it possible to identify the point of departure and evaluate progress through maturity levels (Software Engineering Institute 1995).

Deloitte (Park and Pavlovsky 2010) use maturity-based models for creating competitive advantage "[...] the goal should be to embed sustainability considerations into a company's strategy and operations in such a way as to enhance business value and derive a competitive advantage" (Mani et al. 2010).

In 1995, the Software Engineering Institute (SEI) at Carnegie Mellon University published the book 'The Capability Maturity Model (CMM): guidelines for improving the software process' focusing on the contribution of improving processes in an organization based on the evolutionary path from immature to mature and disciplined processes. Since then, various CMMs have evolved and have been developed for multiple fields and applications that are inspired by CMM structure and processes (Software Engineering Institute 1995; Veleva and Ellenbecker 2001; Veleva et al. 2001; Pinheiro de Lima et al. 2012).

The conceptual model presented in this chapter that supports sustainability indicators generation is based on the extended view of TBL that includes Governance, correlated to value chain activities, and organized according to a maturity-based model. Producing sustainability indicators results in a 'meta data' set defined by Neely et al. (1997) framework. Figure 11.1 summarizes the conceptual framework that theoretically supports the presented research.

Information presented in Fig. 11.1 follows a simple input/process/output structure, defining as input: business strategy definitions, sustainability scope (economic, environmental, social, and governance), value chain scope, and normative documents and standards scope. Process defines how the inputs could be synthesized in a coherent set of measures, transforming normative documents and standards requisites into measure specifications. Output organizes information for publishing sustainability measures and as some graphics to represent relationships among the designed measures and TBL dimensions, value chain activities, and maturity levels.

Defined the conceptual structure, it is possible to present the elements that guide research design.

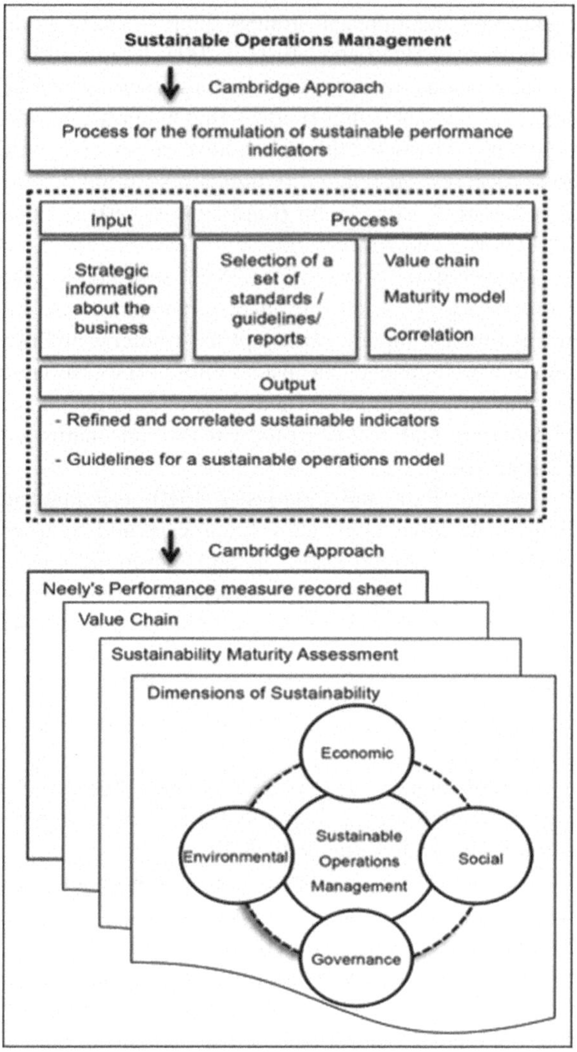

Fig. 11.1 Conceptual model for sustainability indicators generation. *Source* Machado et al. (2012)

11.3 Integrated Sustainability Indicators Development

It is important to clarify that the present study is focused on presenting and discussing the results of a process application for generating a set of integrated sustainability indicators. However, it was designed as software for supporting the proposed process.

The conceptual model is based on best practices for managing sustainability described in sustainability standards, normative documents and report guidelines, international sustainability survey reports, sustainable operations management models, and performance measurement systems frameworks. The application model was developed based on the 'Cambridge Process Approach' (Platts 1993) and it was improved and refined based on the BPM cycle (Baldam et al. 2007). The process for generating sustainability indicators was developed using the BPMN representation standard and operationalized through a BPMS system called Bonita®.

The generic conceptual framework presented in Fig. 11.1 and 'instanced' by Fig. 11.6 is generic and could be applied to any company with a formal sustainability management system. Process prototype and software development resulted in (Machado et al. 2012):

- Input: Sustainability diagnostic procedure survey based on 25 questions related to TBL dimensions and governance structures and processes.
- Processing: Sustainability indicators database generation based on 14 documents: 6 sustainability-related standards—OHSAS 18001, SA 8000, ISO 26000, ISO 14001, ISO 50001, and ISO 9001; 4 normative sustainability guidelines—Global Compact, agenda 21, United Nations Millennium Development Goals, and AA 1000 Accountability; and four sustainability report guidelines—GRI, ISE (BMF&FBOVESPA), Ethos Institute Report, and FNQ Quality Award criteria. Indicators are organized into five maturity levels.
- Output: Sustainability indicators information recording sheet. TBL + Governance report. Value chain report. Maturity levels report. Strategic guidelines report.

Two critical features of the sustainability indicators process/software prototype must be described. The first challenge is to transform sustainability normative framework requisites into indicators.

A process was developed to transform requisites into indicators based on:

- To attend to requisites demands and to allow its management.
- To establish a standard procedure for converting requisites into indicators.
- To have traceability in terms of process information flux.
- To enhance liability and consistency through sustainability indicators database.

The process representing sustainability normative documents requisites conversion into sustainability indicators is presented in Fig. 11.2. (Hundzinski et al. 2013).

If validated from operational and strategic perspectives, a requisite will only be converted into an indicator. Test criteria could be stated as follows (Kennerley and Neely 2003):

- Consistency: To be based on actions that requisite look for development.
- Accuracy: To guarantee requisite compliance through measurement and assessment.
- 'Updates': To create a capability of being continuously updated.
- Traceability: Trace data origin and connect it to the organization's target and goals.

The process ends when the activity 'identification of possible improvements to be generated after model application' reaches an acceptable quality level; that is, the process is repeated through multiple refinements and improvements until indicator validation. The numerous and successive classes are related to:

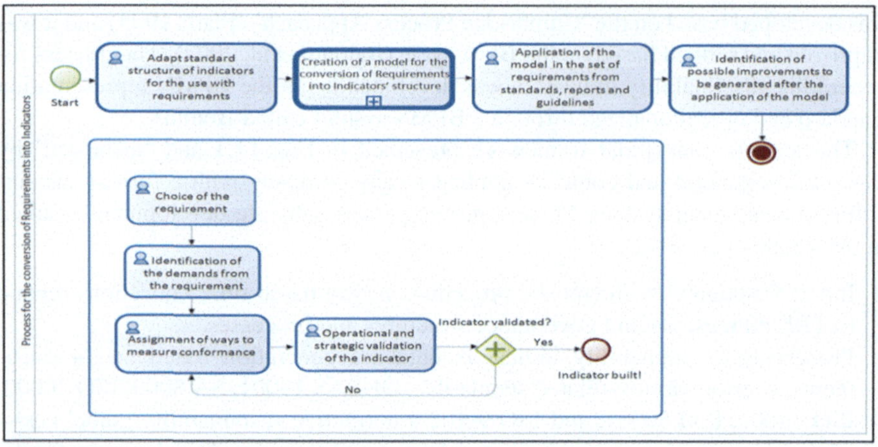

Fig. 11.2 Requisites conversion into sustainability indicators process. *Source* Hundzinski et al. (2013)

- Indicators assessment to assure coherence among them and the requisites related to.
- Value chain and maturity levels categorization improvement process.
- Indicators application classification.

Sustainability indicators follow an information structure inspired by Neely et al. (2002), in which several fields that form integrated information sets define a performance measure (Fig. 11.3).

The second challenge for designing and implementing the sustainability indicators generation process is to correlate indicators. Sustainability normative documents have several points in common, and the proposed process intends to identify them and obtain a lean set of sustainability indicators. For this task, content analysis techniques were applied to the normative sustainability framework to identify similar requisites. Coding procedures were used for producing 'comparative' tags, and they are the critical element for correlating sustainability indicators. Figure 11.4 shows the process of tag creation.

Each activity described in Fig. 11.4 is an iterative process that follows improvement and refinement cycles. Quality criteria established by sustainability experts define the tag creation process ends.

Table 11.1 shows the normative sustainability documents integrated into the process/software prototype. The presented documents were analyzed to transform requisites into indicators, and after that, a correlation algorithm was applied to the sustainability indicators database.

Standards glossary and content helped construct a tag database related to indicators database, being the element that defines indicators similarity.

Tags allow the development of a correlation algorithm. Sustainability indicators content correlation is based on correlation and application degrees.

Indicator record	Existing indicator
Standard	
Requirement	
Indicator	
Description	
Category	
Theme	
Sub-theme	
Dimension	
Value Chain	
Maturity Level	
Type	
Correlation	
Propose	
Refers to	
Target	
Formula / Metrics	
Frequency of measurement	
Frequency of reviews	
Who measures?	
Data source	
Who is responsible for the measurement?	
What do they do?	
Who act on the data?	
What do they do?	
Notes and commentary	

Fig. 11.3 Sustainability indicators record sheet. *Source* Neely et al. (2002)

The correlation degree measure is applied when two texts are compared to know how similar they are. The values range from 0 to 100%, depending on the correlation degree between the texts.

Stating:

- Text A = Text that is being compared to the other texts group.
- Text Bn = Group text where similarity comparison is being carried out.

And considering:

- NTAtA = Number of Tags assigned to text A.
- NtB = Number of Tags assigned to text B.
- NtE = Number of equivalents Tags between the texts.
- NtT = total Tags number of the texts.

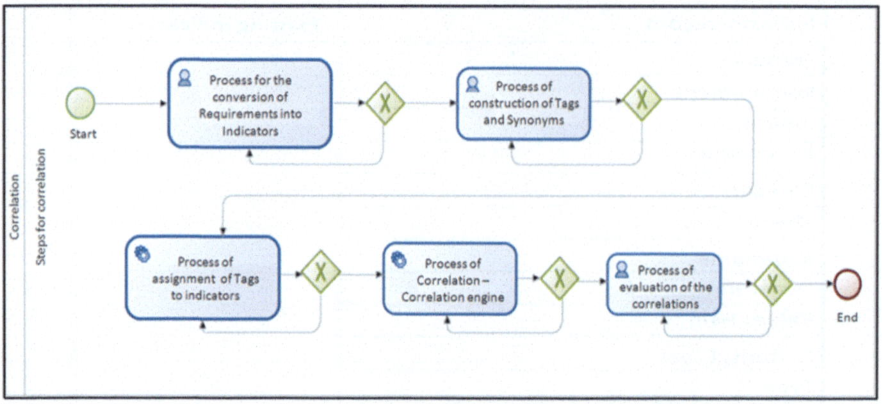

Fig. 11.4 Tag creation process and procedures. *Source* Kluska et al. (2013)

Table 11.1 Studied sustainability normative documents

Document	Characteristics
NBR ISO 9001	Quality management system
NBR ISO 14001	Environmental management system
NBR ISO 16001	Social responsibility management system
NBR ISO 26000	Guidelines about social responsibility
NBR ISO 50001	Energy management system
OHSAS 18001	Health and occupational safety system
SA 8000	Social responsibility
Brazilian agenda 21	Sustainable responsibility
Global compact	Human rights, work relationships, environment and measures against corruption
AA 1000 accountability	Social responsibility management
Ethos indicators	Corporate social responsibility
GRI (Global reporting initiative)	Sustainability reports
ISE—BM&FBOVESPA	Corporate sustainability, based in economic efficiency, environmental balance, social justice, and corporate governance
Critérios Rumo à Excelência—FNQ	Organizational management, aiming to obtain excellence in performance

$$\text{NtT} = \text{NtA} + \text{NtB} \tag{11.1}$$

I am PtE as the percentage of the equivalent Tags number, by the total Tags assigned a number to the two indicators.

$$\text{``PtE} = \text{''} \ \text{``NtE} \times 2\text{''} / \text{``NtT''} \ \text{``} \times 100\ \%\text{''} \tag{11.2}$$

PtE represents the equivalent tags intercession between the texts, Text A ∩ Text B.

Application degree is used to verify the text application to another text. This correlation analysis model assigns values ranging from 0 to 100% depending on the text application degree.

Stating:

- Text A = Text that is being compared to the other group texts.
- Text Bn = Group text where similarity comparison is being undergone, and the application degree is calculated.

And considering:

- NTAtA = Number of Tags assigned to text A.
- NtE = Number of equivalents Tags between the texts.
- PAt = Percentage of text application before another text.

$$\mathrm{PAt} = (\mathrm{NtA} \times 2)/(\mathrm{NtA} + \mathrm{NtE}) \times 100\% \tag{11.3}$$

The PAt value represents the text A equivalent tags intercession with the equivalent tags between the texts, Text A ∩ (Text A ∩ Text B).

Figure 11.5 shows the correlation process based on content analysis. Some elements used for the operating correlation algorithm are shown, including database use. This feature is critical since all correlation scripts and assignments work with simultaneous playback loops. The storage architecture offers this possibility.

After presenting the main characteristic of the process/software prototype, it is possible to discuss its application and test in a case.

Company ZZ is a global leader in sustainable forest product development. The company has been operating in this business for the last five decades. Its mission is founded on sustainably creating forest product value, integrating high standards in forestry production with efficiency in industrial transformation for its global distribution by its customer's needs.

The test conducted in Company ZZ was planned according to the following activities:

- Sustainability diagnostic survey application.
- Inputs formalization to orient the process of generating sustainability indicators.
- Assessment of sustainability indicators generated a list of present and desired situations.
- Sustainability indicators assessment report workshop.
- Workshop for completing sustainability indicators record information sheet.

It is a mature company that defines its multiple businesses based on sustainability principles. Although it does not yet have formal Corporate Social Responsibility (CSR) policies, it attends to a complete set of requirements related to TBL and Governance best practices. The company obtained 95 points on a 100 points scale sustainability diagnostic survey, one of the first activities in the prototype test.

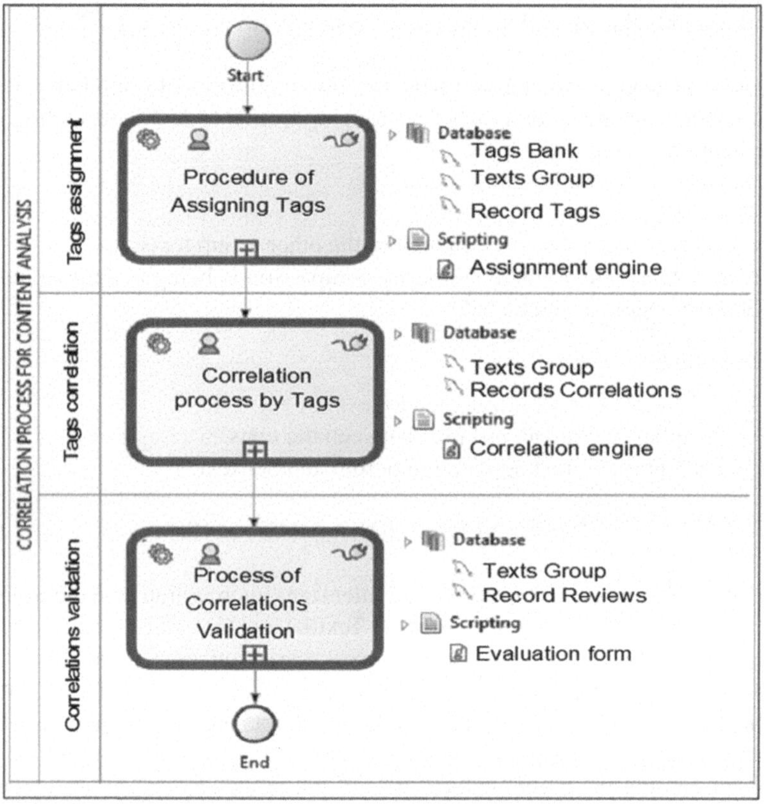

Fig. 11.5 Indicators correlation process. *Source* Kluska et al. (2013)

After completing the sustainability diagnosis survey, it was possible to define inputs for the prototype test. This phase works in a logic "As Is"—"To Be." The inputs are described by:

- Scope: The company must select a product, process, or service perspective. This information helps researchers in the coaching process. Company ZZ chose a process perspective.
- Extended TBL dimensions: Depending on company strategy and maturity level, it could choose to work with the entire set of social, environmental, economic, and governance dimensions or a subset. Company ZZ's exercise was focused on social and ecological dimensions.
- Value chain activities: Inbound logistics, operations, outbound logistics, marketing, sales, after-sales, infrastructure (technology and organizational), human resources management, and procurement define value chain activities, and the company could select the ones to be studied. Company ZZ selected infrastructure and operations to be studied.

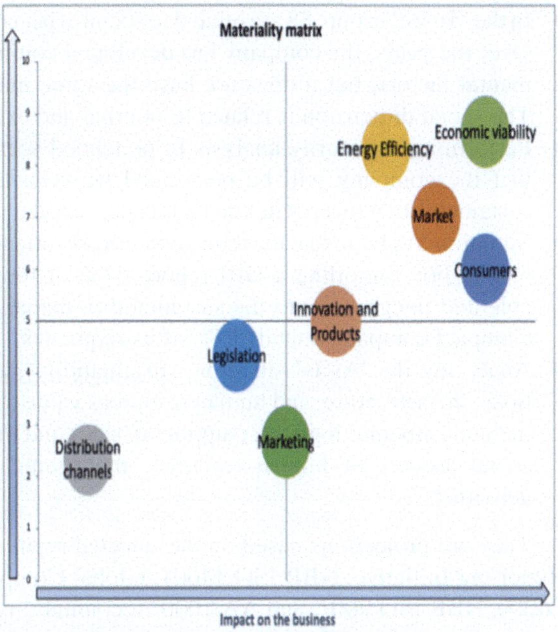

Fig. 11.6 Materiality matrix. *Source* Carpejani et al. (2017)

- Present situation (As Is): Company present situation in implementing normative sustainability framework. ISO 9001, ISO 14001, OHSAS 18001, GRI, and Global Compact define its current normative model.
- Future situation (To Be): The company's desired position in implementing a normative sustainability framework. Company selected ISO 26000 and AA 1000 Accountability to be included in its sustainability normative model.

Defined the inputs, a list of 176 sustainability indicators were generated. The company has 76 hands as implemented in its sustainability operations management system; that is, 100 sustainability indicators could be integrated into its operations management system.

Some general results could be summarized from the case application:

- A sustainability diagnostic survey identified opportunities for integrating new sustainability indicators from a social perspective.
- A list of 176 elements forms the entire set of proposed sustainability indicators.
- Seventy-six sustainability indicators are already implemented, forming the 'As Is' group. It is intended to implement more than 100 indicators that include the group 'To Be.'
- 'As Is' group comprises 40 indicators related to social aspects and 36 environmental ones. The group is balanced.

- In the 'To Be' group, 88 are related to social aspects and 12 to environmental issues. Over the years, the company has developed competences for managing environmental factors, but it does not have the same maturity in the social dimension. The social dimension is related to internal and external themes.
- Performing a maturity analysis in pretended scenario 'To Be,' it could be seen that the company will be positioned between levels 4 and 5, indicating that sustainability concepts define its business model and strategy.
- Situation 'As Is' reveal that company already attended 22% of ISO 26000 requirements. For compiling a GRI report, 51% of the required information is available and integrated into the sustainability management system. For UN Global Compact company attends 57% of its requisites.
- Analyzing the 'As Is' situation, sustainability indicators are oriented to operations, infrastructure, and human resources value chain activities. 'To Be' scenario includes inbound logistics (suppliers), sales and marketing, and intensify internal social aspects of human resources management and operations value chain activities.

They are projections based on the selected regulatory framework: GRI—Global Reporting Initiative, NBR ISO 14001, Global Compact, OHSAS 18001, NBR ISO 26000, NBR ISO 9001, and AA 1000 Accountability.

Scenario 1 is based on the seven selected documents and does not use any TBL aspects or value chain activities restrictions. It could be taken as the complete sustainability performance measurement system. Their broad input forms a set of 304 indicators. Table 11.2 shows some numbers analyzing the correlation among the indicators.

Table 11.2 Scenario 1: sustainability performance measurement system

Correlation algorithm output	
92,416	Total
Complete scenario considering all extended TBL dimensions and value chain activities	
304	Total number of possible indicators
154	Indicators that are correlated to another hand in the selected regulatory framework
150	Indicators that are **not** correlated to another hand in the selected regulatory framework
187	Indicators that are correlated by another indicator in the selected regulatory framework
117	Indicators that are **not** correlated with another indicator in the selected regulatory framework
194	A minimum set of indicators that covers the selected regulatory framework
Indicators improvement analysis	
36,20%	Improvement rate
51%	Number of indicators that have any correlation
61,50%	Number of indicators represented in the regulatory framework

It could be seen that 154 indicators are the set that 'points to' at least one indicator in the sustainability regulatory database; that is, in the worst situation, it could be possible to have 77 indicators to manage all requisites contained in these indicators. One hundred and seventeen represent the indicators not pointed out by any other indicators in the sustainability regulatory database. Suppose it is summed 117 as a minimum set of individual indicators and 77 as the lean set of correlated indicators. In that case, the worst situation forms a collection of 194 indicators for managing the integrated sustainability system.

Scenario 2 is formed by Company ZZ's choices regarding TBL aspects and value chain activities. The total number of possible indicators decreased to 176, and Table 11.3 shows the improvements that could be obtained through the correlation algorithm application.

Scenarios 1 and 2 show improvement potential in indicators numbers by running a correlation algorithm. It is clear for these two cases that an improvement rate of 30%, but if an individual indicator analysis is performed, the improvement rate tends to be better.

Indicators analysis provided a systemic and integrated view of the entire system for managing sustainability, providing collective learning about connections, interfaces, and integration that could be implemented in processes and systems used for managing sustainability aspects and issues.

Table 11.3 Scenario 2: sustainability performance measurement system

Correlation algorithm output	
30,976	Total
Reduced scenario considering a selected number of extended TBL dimensions and value chain activities	
176	Total number of possible indicators
85	Indicators that are correlated to another indicator in the selected regulatory framework
91	Indicators that are **not** correlated to another indicator in the selected regulatory framework
99	Indicators that are correlated by another indicator in the selected regulatory framework
77	Indicators that are **not** correlated with another indicator in the selected regulatory framework
120	A minimum set of indicators that covers the selected regulatory framework
Indicators improvement analysis	
32,1%	Improvement rate
48%	Number of indicators that have any correlation
56,3%	Number of indicators represented in the regulatory framework

11.4 Lessons Learned and Future Trends

Results show that the developed process could generate an integrated and 'lean' set of sustainability indicators. Improvements opportunities are identified, especially those related to the sustainability diagnostic survey and the indicators correlation model. The process could allow company ZZ to simulate different conditions in terms of scope to develop its sustainability indicators, creating conditions for learning and improvement.

In terms of methodology, procedures developed follow systemic requirements for action research conduction, and the multiple and successive refinement and learning loops or cycles assured results quality. It is essential to observe that the BPM cycle model with its intrinsic improvement characteristic is crucial for indicator design. Its presence is verified at different levels from requisites, tags, and indicators generation.

Companies that are qualified to apply the process could reach: (i) Competence for identifying indicators for new certification implementation processes; (ii) Shorten the time required for implementing new elements in sustainability regulatory framework (standards, reference documents, and report guidelines); (iii) Low cost for implementing a new process and certification; (iv) Simplification in maintaining and actualizing the sustainability regulatory framework; (v) Integrated management for sustainable operations.

These clear benefits contribute to a lean process in managing sustainability, and the management is essentially defined through measurement.

According to Yudhanti and Tjahjadi (2021), the company's size affects corporate social responsibility disclosure. This result indicates that companies with significant assets tend to be better able to conduct and transmit social responsibility information. Companies should be aware that disclosing sufficient information to the public, of course, requires investment. Therefore, companies need to be supported with their resources, as reflected in the assets owned by the company. The practice of social responsibility requires several steps, from the planning stage to the delivery of information.

To prepare a Sustainability Report based on the Global Report Initiative (GRI), it is necessary to follow the so-called fundamentals. GRI is an independent international organization that helps companies and other organizations take responsibility for their impacts by providing them with the common global language to communicate these impacts. They give the world's most widely used standards for sustainability reporting, the so-called GRI Standards. The fundamentals are a kind of philosophy of GRI, being "the starting point for the use of the GRI Standard." They are responsible for defining the content and quality of the report. The GRI Standards-based sustainability report should provide a balanced and sensible description of an organization's positive and negative contributions to sustainable engagement. The information provided through the sustainability report should allow the organization's internal and external stakeholders to form opinions and make informed decisions about the organization's contributions to sustainable development (GRI 101 2018).

According to GRI 101 (2018), a sustainability report is structured around principles for defining the content of the information and principles for determining the quality of the report. The content principles cover the following scope: stakeholder inclusion, sustainability context, materiality, and completeness. On the other hand, the principles of quality must guarantee the report: accuracy, balance, clarity, comparability, reliability, and timing.

The G4 guideline of the current GRI 101 standard (2018) emphasizes material principles or materiality to select and define the indicators used throughout the report. Until the G4 version, the material topics were synonymous with material aspects; however, in the Norm, the word "aspects" was no longer used because it was not clear. The official version calls material topics (or material themes, in direct translation of Spanish) or can be called 'material questions.' A material topic should reflect the significant economic, environmental, and social impacts of an organization producing the report, that is, topics that substantially influence stakeholder assessments and decisions. To have its Sustainability Report recognized by GRI to comply with the standards, the organization must inform its relevant topics.

When the organization classifies a topic as material, it should inform how it is its management approach; that is, it is not enough to tell; it is necessary to show the limits and the way of management. According to Jones et al. (2016), describing the boundaries of each impact is critical to the consolidation of the report. The initial description of each indicator, regardless of its series (200 for economic indicators, 300 for environmental, and 400 for social indicators), shows the importance of the obligation to report. According to Calabrese et al. (2016), the reports must have the appropriate levels of detail to justify the criterion adopted in the choice of material topics.

According to GRI 101 (2018), an organization to select its material topics should observe the indicator books 200, 300, and 400. However, the organization can include the new topic-related indicator if a found material topic is not covered. Also, if any material topic does not cover a series (200, 300, or 400), it does not need to be reported. For example, a company may include the contribution to the 17 SDGs in its topics, which goes beyond the perspective of the 200 series notebooks: standards related to economic issues; 300 series: standards about environmental problems; and series 400: standards related to social topics. It may also include indicators related to digital transformation that contribute directly or indirectly to the hands in the 200, 300, or 400 notebooks. It is essential to be in mind the fundamentals of material topics in their objective contribution to the business strategy and the realization of the company's mission.

Material topics represent the subset of materiality. The critical questions are grouped and formed materiality, and according to GRI 101 (2018), materiality is the threshold from which the questions become expressive enough to be reported. Beyond this threshold, not all material topics have the same importance, as the report's emphasis should reflect its relative priority. Materiality should include relevant secular themes capable of reflecting an organization's economic, environmental, and social impacts and thus confronting its priority for stakeholders. The results can be both negative and positive perspectives. Each material theme has different importance in the context of materiality, so it is expected that the company shows the relative importance of secular topics in the reports.

The relative importance makes materiality provides stakeholders with only important, complete information that should assess sustainable business performance. Questions should be classified according to a scale of importance, from most minor to most relevant. Relevance is an essential part of the principle of materiality, as organizations should focus only on necessary activities for the business and the external public. Applicability, therefore, provides greater transparency of sustainable practices for society. Materiality should be explained in text reports and visually through a materiality matrix (Whitehead 2017; Jones et al. 2016; Calabrese et al. 2016).

To Bellantuono et al. (2016) and Jones et al. (2016), the materiality matrix is an effective way for the organization to demonstrate the relevant material topics of its business. Building the matrix is a significant step to correctly identifying the material principles of an organization. Briefly, the matrix is the graphic expression of materiality results, synchronizing material topics from the company's perspective and its stakeholders. The result is the visual representation of the prioritization of material topics, as illustrated in Fig. 11.6.

The materiality matrix is the graphical expression of materiality results, synchronizing material topics from the company's perspective and its stakeholders. The axis of abscissas x is responsible for the importance of material issues, described in the economic, environmental, and social perspectives. At the same time, the axis of they-ordered represents the influence of stakeholders on these issues. In GRI standards, there is no specific information on how to prioritize, a kind of approach, or methodology. However, the company must ensure a set of procedures in its elaboration, including interest groups, legislation, and topics relevant to the sector of activity.

Carpejani et al. (2017) observe that the structure of the materiality matrix is similar, from the visual perspective to the matrix of importance and performance. However, the operationalization and analysis are somewhat different. The matrix analysis suggests that each topic has its relative importance and, consecutively, the issues in the upper right quadrant are the most important. Thus, the items should be prioritized in the performance measurement and sustainability report.

Acknowledgements The authors also wish to thank CNPq (National Council of Technological and Scientific Development) for supporting the research project through grant 317150/2021-9.

References

Aras G, Crowther D (2008) Governance and sustainability: an investigation into the relationship between corporate governance and corporate sustainability. Manag Decis 46(3):433–448

Azapagic A, Perdan S (2000) Indicators of sustainable development for industry: a general framework. Process Saf Environ Prot 78(4):243–261

Baldam R, Abreu M, Hilst S, Pereira H (2007) Business process management—BPM, Erica, São Paulo (in Portuguese)
Bellantuono N, Pontrandolfo P, Scozzi B (2016) Capturing the stakeholders view in sustainability reporting: a novel approach. Sustainability 8(4):379
Cabezas H, Pawlowski CW, Mayer AL, Hoagland NT (2003) Sustainability: ecological, social, economic, technological, and systems perspectives. Clean Technol Environ Policy 5(3–4):167–180
Calabrese A, Costa R, Levialdi N, Menichini T (2016) A fuzzy analytic hierarchy process method to support materiality assessment in sustainability reporting. J Clean Prod 121:248–264
Carpejani P, Pinheiro de Lima E, Machado CG (2017) What is materiality in GRI G4? A discussion and agenda for future research. In: Proceedings of the 6th international workshop on advances in cleaner production, p 10
Castka P, Balzarova MA (2007) A critical look on quality through CSR lenses: key challenges stemming from the development of ISO 26000. Int J Q Reliab Manage 24(7):738–752
Epstein MJ, Roy MJ (2001) Sustainability in action: identifying and measuring the key performance drivers. Long Range Plan 34(5):585–604
Gold S, Seuring S, Beske P (2010) Sustainable supply chain management and inter-organizational resources: a literature review. Corp Soc Responsib Environ Manag 17(4):230–245
Golini R, Longoni A, Cagliano R (2014) Developing sustainability in global manufacturing networks: the role of site competence on sustainability performance. Int J Prod Econ 147(Part B):448–459
GRI 101 (2018) Fundamentos. GRI, Amsterdam
Hart SL, Milstein MB (2003) Creating sustainable value. Acad Manag Exec 17(2):57–69
Hasna AM (2010) Sustainability classifications in engineering: discipline and approach. Int J Sustain Eng 3(4):258–276
Hundzinski LN, Pinheiro de Lima E, Gouvea Da Costa SE, Machado CG, Cestari JMAP, Kluska RA (2013) Sustainability standards and guidelines requirements for integrated management. Proceedings of 22nd international conference on production research—ICPR 22, Foz do Iguaçu, Brazil, pp 1–9
Hutchins MJ, Sutherland JW (2008) An exploration of measures of social sustainability and their application to supply chain decisions. J Clean Prod 16(15):1688–1698
Isaksson R, Steimle U (2009) What does GRI-reporting tell us about corporate sustainability? TQM J 21(2):168–181
Jones P, Comfort D, Hillier D (2016) Managing materiality: a preliminary examination of the adoption of the new GRI G4 guidelines on materiality within the business community. J Public Aff 16(3):222–230
Keeble J, Topiol S, Berkeley S (2003) Using indicators to measure sustainability performance at a corporate and project level. J Bus Ethics 44(2–3):149–158
Kennerley M, Neely A (2003) Measuring performance in a changing business environment. Int J Oper Prod Manag 23(2):213–229
Kiron D, Kruschwitz N, Hannaes K, Reeves M, Velken IVS, Audretsch M (2012) Sustainability nears a tipping point. MIT Sloan Manag Rev 53(2):69–74
Kleindorfer PR, Singhal K, Van Wassenhove LK (2005) Sustainable operations management. Prod Oper Manag 14(4):482–492
Kluska RA, Pinheiro de Lima E, Gouvea Da Costa SE, Machado CG, Cestari JMAP, Hundzinski LN (2013) Correlation process in content analysis for a BPM modeling project. Proceedings of 22nd international conference on production research—ICPR vol 22. Foz do Iguaçu, Brazil, pp 1–8
Krajnc D, Glavič P (2005) How to compare companies on relevant dimensions of sustainability. Ecol Econ 55(4):551–563
Labuschagne C, Brent AC, Erck RPG (2005) Assessing the sustainability performances of industries. J Clean Prod 13(4):373–385

Leppelt T, Foerstl K, Reuter C, Hartmann E (2013) Sustainability management beyond organizational boundaries–sustainable supplier relationship management in the chemical industry. J Clean Prod 56(1):94–102

Looy AV, De Backer M, Poels G, Snoeck M (2013) Choosing the right business process maturity model. Inf Manag 50(7):466–488

Louette A (2008) Compendium for sustainability: social and environmental responsibility management tools, 2nd edn. Willis Harman House/Antakarana, São Paulo

Maas S, Reniers G (2014) Development of a CSR model for practice: connecting five inherent areas of sustainable business. J Clean Prod 64(1):104–114

Machado CG, Pinheiro de Lima E, Gouvea Da Costa SE, Cestari JMAP, Kluska RA, Hundzinski LN (2012) Indicators formulation process for sustainable operations management. Proceedings of international conference on production research—ICPR Americas 2012. Santiago, Chile, pp 1–7

Mani M, Lyons K, Sriram R (2010) Developing a sustainability manufacturing maturity model. Proceedings of IMS 2020 summer school on sustainable manufacturing, May 26–28. Zurich, Swiss, pp 311–321

McCartney D (2009) Designing corporate sustainable development indicators: reflections on a process. Environ Qual Manage 19(1):31–42

Meadows D (1998) Indicators and information systems for sustainable development. The Sustainability Institute, Hartland Four Corners

Mori K, Christodoulou A (2012) Review of sustainability indices and indicators: towards a new City Sustainability Index (CSI). Environ Impact Assess Rev 32(1):94–106

Neely AD, Bourne MCS, Mills JF, Platts KW, Richards H (2002) Strategy and performance: getting the measure of your business. Cambridge University Press, Cambridge, UK

Neely AD, Richards H, Mills JF, Platts KW, Bourne MCS (1997) Designing performance measures: a structured approach. Int J Oper Prod Manag 17(11):1131–1152

Nidumolu R, Prahalad CK, Rangaswami MR (2009) Why sustainability is now the key driver of innovation? Harv Bus Rev 87(9):3–10

Parisi C (2013) The impact of organisational alignment on the effectiveness of firms sustainability strategic performance measurement systems: an empirical analysis. J Manage Governance 17(1):71–97

Park C, Pavlovsky K (2010) Sustainability in business today: a cross-industry view. Deloitte Development LLC, New York

Pinheiro de Lima E, Gouvea da Costa SE, Machado CG, Manfrim PM (2012) Sustainable operations strategy: theoretical frameworks evolution. Proceedings of EUROMA, 2012, 4th world conference on production and operations management. Amsterdam, Netherlands, pp 1–10

Platts KW (1993) A process approach to researching manufacturing strategy. Int J Oper Prod Manag 13(8):4–17

Porter M, Kramer M (2006) The link between competitive advantage and corporate social responsibility. Harv Bus Rev 84(12):78–91

Reuter C, Goebel P, Foerstl K (2012) The impact of stakeholder orientation on sustainability and cost prevalence in supplier selection decisions. J Purch Supply Manag 18(4):270–281

Schrettle S, Hinz A, Scherrer-Rathje M, Friedli T (2014) Turning sustainability into action: explaining firms sustainability efforts and their impact on firm performance. Int J Prod Econ 147(Part A):73–84

Software Engineering Institute (1995) The capability maturity model: guidelines for improving the software process. Addison-Wesley, Reading

Tate WL, Ellram LM, Kirchoff JONF (2010) Corporate social responsibility reports: a thematic analysis related to supply chain management. J Supply Chain Manag 46(1):19–44

Ueda K, Takenaka T, Váncz J, Monostori L (2009) Value creation and decision-making in sustainable society. CIRP Ann Manuf Technol 58(2):681–700

van Bommel HWM (2011) A conceptual framework for analyzing sustainability strategies in industrial supply networks from an innovation perspective. J Clean Prod 19(8):895–904

Veleva V, Ellenbecker M (2001) Indicators of sustainable production: framework and methodology. J Clean Prod 9(6):519–549

Veleva V, Hart M, Greiner T, Crumbley C (2001) Indicators of sustainable production. J Clean Prod 9(5):447–452

WCED (1987) Our common future. Oxford University Press, Oxford

Whitehead J (2017) Prioritizing sustainability indicators: using materiality analysis to guide sustainability assessment and strategy. Bus Strateg Environ 26:399–412

Wilkinson A, Hill M, Gollan P (2001) The sustainability debate. Int J Oper Prod Manag 2(12):1492–1502

Yudhanti CBH, Tjahjadi B (2021) Drivers of social responsibility disclosure: the moderation of the president director's busyness and political connections. Asian J Account Res 6(3):335–34 (2021)

Edson Pinheiro De Lima Received his Ph.D. in Industrial Engineering from Federal University of Santa Catarina (UFSC), his master's degree in electrical engineering (Automation) from University of Campinas (UNICAMP) and his Bachelor of Science in Electrical Engineering from Federal University of Technology—Parana (UTFPR). He is a Professor at the Industrial and Systems Engineering of Federal University of Technology—Parana (UTFPR), Brazil. During his Ph.D. he spent 12 months as a Visiting Researcher at the Logistics and Production Engineering Laboratory of the Production Engineering, Business and Statistics Department of the Industrial Engineering School at Polytechnic University of Madrid, Spain. Professor Edson Pinheiro is being over the years a Visiting Scholar in the Warwick Business School (2007), Polytechnic University of Madri (2010) and ISE at Virginia Tech (2017–2018). His current research interests include sustainable operations, operations strategy, performance measurement systems and performance analytics.

Pablo Carpejani Ph.D. candidate in Industrial and Systems Engineering at Pontifical Catholic University of Parana (PUCPR), his master's degree in industrial and systems engineering from Pontifical Catholic University of Parana (PUCPR), and his Bachelor of Science in Business Administration from Pontifical Catholic University of Parana (PUCPR). He is an Assistant Professor at the Business of Pontifical Catholic University of Parana (PUCPR), Brazil. He participated in sustainability reporting processes, from measuring performance, creating the materiality process, and publishing the report. He is currently coordinator of the MBA in corporate sustainability at PUCPR. His current research interests include sustainable operations, GRI reports, ESG models, and sustainability materiality.

Carla Gonçalves Machado Received his Ph.D. in Industrial and Systems Engineering from Pontifical Catholic University of Parana (PUCPR), his master's degree in industrial engineering from State University of Sao Paulo (UNESP) and his Bachelor of Science in Business/Corporate Communication—Public Relations from State University of Sao Paulo (UNESP). Between 2018 and 2020, Carla was Postdoctoral Researcher and Researcher at the Division of Supply & Operations Management involved in research projects on Sustainability, Digitalization and Supply & Operations Management, also teaching in Master courses—Operations Strategy; Manufacturing Strategies; Business Marketing and Purchasing. She is an Assistant Professor in Business Administration at Jönköping International Business School, Jönköping University, Sweden. She is Director of Societal Engagement at Jönköping International Business School. Her current research interests include sustainable operations management, digitalization, and supply chain management.

Sergio E. Gouvea Da Costa Received his Ph.D. in Industrial Engineering from University of São Paulo (USP), his master's degree in electrical engineering (Automation) from University of Campinas (UNICAMP) and his Bachelor of Science in Electrical Engineering from Federal University of Technology—Parana (UTFPR). During his doctorate, he carried out research at the Institute for Manufacturing da University of Cambridge, England (2000–2001). He did Postdoctoral work at the Edward P. Fitts Department of Industrial and Systems Engineering at North Carolina State University, USA (2009–2010). He is an Associate Professor at the Industrial and Systems Engineering of Federal University of Technology—Parana (UTFPR), Brazil. His current research interests include Energy Management, Operations Strategy and MCDM/A.

Chapter 12
Analysis of Ergonomic Risks in Remote Work

Letícia De Oliveira Gago Ramos De Souza and Eduardo De Oliveira

Abstract This chapter explores the impact of Corporate Social Responsibility (CSR) and ergonomics on the financial performance of organizations. It evaluates the transition to the telework model during the COVID-19 pandemic, considering ergonomic precautions and their effect on workers' quality of life. The study uses a quantitative and descriptive research method, employing an online questionnaire for Brazilian teleworkers. The responses will reveal the guidance provided by employers and the precautions taken in the telework environment.

Keywords Ergonomics · Sustainability · Social · Remote work · Pandemic

12.1 Introduction

In the current societal landscape, two prominent themes emerge: ESG (Environmental, Social, Governance), which evolves from the concept of sustainability and aims to balance economic, environmental, and social objectives; and the issue of remote work, which arises in the wake of technological revolutions which gains momentum in the context of COVID-19. There's a need for analysis of ergonomic risks in this type of work. This is especially crucial post-COVID-19, where remote work actions can be scrutinized and planned to prevent the gains from being offset by employee health risks.

Within the ESG context, particularly concerning labor and consumer markets and society, this chapter casts a new light on employment relationships and organizations' social objectives regarding remote work analysis, examining how organizations manage their workforce. Despite advancements, economic emphasis persists, while social challenges are often overlooked, especially in the remote work context.

L. D. O. G. R. De Souza · E. De Oliveira (✉)
Department of Production Engineering, Faculty of Engineering, Universidade Federal Fluminense, Niterói, Brazil
e-mail: eoliveira@id.uff.br

M. J. Meiriño and O. L. G. Quelhas (eds.), *Critical Factors Associated with Environmental, Social and Governance Management in Organizations*, Accounting, Finance, Sustainability, Governance & Fraud: Theory and Application, https://doi.org/10.1007/978-981-96-1725-8_12

Through bibliometric study in this chapter, the increasing importance of telecommuting is revealed, highlighting ergonomic challenges and the balance between professional and personal life. Case studies illustrate how companies can swiftly adapt, implement security measures, and maintain high levels of satisfaction during challenging periods.

12.2 Role of the Chapter in the Structure of the Book

The theme Environmental, Social and Corporate Governance (ESG) is highlighted in the market and society and is used by investors to analyze potential investments. However, the concern with environmental, social, and governance aspects is not incipient since it is perceived as an evolution of the sustainability tripod created by John Elkington. As previously mentioned, the idea of corporate sustainability proposes a new economy, in which companies, to be sustainable, should balance their objectives, being "financially viable, environmentally responsible and socially fair" (Elkington 1994). As said by Elkington 25 years later, even though the triple bottom line had made evolve the question of better socio-environmental outcomes, there is no change in capitalism because organizations continue to focus more on the economic aspect (Elkigton 2018).

The social aspect on the other hand is considered the forgotten aspect of the sustainability theme (OPP 2017), and one of the reasons is the difficulty in defining indicators to measure the social aspect in sustainability (Hutchins et al. 2019). According to the theme of this book, which is related to Environmental, Social, and Governance (ESG), this chapter is linked to the social aspect and area as it will present the ergonomic risks of remote work, when it is taken without the necessary measures and precautions, either for the worker or for the work environment, and its potential damages to companies and for employees.

12.3 History of the Development of the Theme

As previously mentioned, the social part of the ESG is sometimes forgotten and to perform the bibliometric research of the study in question, we used both the Scopus and Web of Science databases, and those searches were made from February 17 to 20, 2022. It is important to highlight that the entire process was divided into two stages: selection of sample articles through research filters and, finally, the final analysis of abstracts of the selected articles, to confirm adherence of an article to the study in question.

To search in those databases, it is necessary to define and set a group of words that have adherence to the theme. So, before the search, the words were delimited as the Boolean logic which was used. That was made to filter and unify the articles found in both databases, the terms used were "Ergonomics AND 'REMOTE WORK' OR telecommuting". Also other search parameters were used as articles published in scientific journals ("articles"), containing the search terms in the title, summary, or keywords in the Scopus database and all fields in the Web of Science database of both academic bases. Additionally, a was used to set the articles found to be from 2019 to the most recent. The result was sixteen articles in the Scopus database and 177 publications of articles in the Web of Science database. Thus, the next step was characterized by the elimination of duplication and adaptation to the theme according to the reading of abstracts, which resulted in seventeen articles in the Scopus database and eighteen articles in the Web of Science database, that is, thirty-five articles in total (Table 12.1).

12.4 Classical Conceptual Structure

According to Alshehhi et al. (2018), competitiveness between companies has increased, bringing changes quickly, and requiring companies to adapt to the new sustainable business model, where they focus on achieving satisfactory financial and economic performance, not in the short term, but the pursuit of a long-term economic, environmental, and social sustainability.

Carlos and Morais (2017) said that a company contributes to sustainable development when they produce environmental, social, and governance benefits, based on the ESG concept.

The social dimension is given by the responsibility that the company needs to demonstrate toward society, including transparent communication through the dissemination of non-financial reports and the inclusion of practices that benefit both the internal and the external community to the organization. This aspect has reference to the workforce, human rights, community, and product responsibility (Soschinski et al. 2019).

Oželienė (2018) posits that the social dimension presents aspects related to employees and labor practices, safety and health, communication with customers, and relationship with communities.

In early 2020, the world witnessed one of the most unusual events of the century, the arrival of a pandemic that led to an immediate shutdown of economic activities around the world (ILO 2020).

With the emergence of the COVID-19 virus, worldwide isolation was determined as a preventive measure and public health containment (Cuerdo-Vilches et al. 2020).

The isolation measures adopted were as: closing schools, social distancing, and closing workplaces because they sought to provide a safe way to prevent the virus from continuing to spread within society and causing a second wave worse than the first (ILO 2020).

Each sector has reacted differently to the situation considering the sociodemographic characteristics, the qualification, and the geographical area or the level of restriction imposed, etc. (Abellán García et al. 2021).

Table 12.1 Summary of the literature on ergonomics and telework

Author	Title
Azmetova, E.A., Vadulina, N.V. (2021)	Features of incidents investigation during remote work
Bellmann, L; Hubler, The (2021)	Working from home, job satisfaction, and work–life balance—robust or heterogeneous links?
Birchmeier, B; Dyck, E; Baker, KP; Buhler, S; Lebert, The (2020)	A compilation of short takes on working from home
Cuerdo-Vilches, T; Navas-Martin, MA; March, S; Oteiza, I (2021)	Adequacy of telework spaces in homes during the lockdown in Madrid, according to socioeconomic factors and home features
Cuerdo-Vilches, T., Navas-Martín, M.A., Oteiza, I. (2021)	Working from home: Is our housing ready?
Danker (Yap; Zalzuli), Ho; Ang (2021)	Surviving work from home: observations from Singapore
Davis, KG; Kotowski, IF; Daniel, D; Gerding, T; Naylor, J; Syck, M (2020)	The home office: Ergonomic lessons from the "New normal"
Dingel, JI; Neiman, B (2020)	How many jobs can be done at home?
Emerson, S., Emerson, K., Fedorczyk, J. (2021)	Computer workstation ergonomics: current evidence for evaluation, corrections, and recommendations for remote evaluation
Feng, ZY; Savani, K (2020)	COVID-19 created a gender gap perceived in work productivity and job satisfaction: implications for dual-career parents working from home
Garcia-Gonzalez, MA; Torrano, F; Garcia-Gonzalez, G (2020)	Analysis of Stress Factors for Female Professors at Online Universities
Gerding, T., Syck, M., Daniel, D., (…), Huston, T.R., Davis, K.G. (2021)	An assessment of ergonomic issues in the home offices of university employees sent home due to the COVID-19 pandemic
Gleason, AM (2021)	Remote monitoring of a work-from-home employee to identify stress: a case report
Goldenberg, I., Denomme, W.J. Lee, J.E.C. (2022)	Pandemic-related challenges and organizational support among personnel in Canada's defense establishment
Jamal, MT; Anwar, I; Khan, NA; Saleem, I (2021)	Work during COVID-19: assessing the influence of job demands and resources on practical and psychological outcomes for employees
Kilcullen, M., Feitosa, J., Salas, E. (2021)	Insights from the virtual team science: rapid deployment during COVID-19
Kuijer, P.P.F.M., van der Pas, J., van der Molen, H.F. (2020)	Work disabling nerve injury at both elbows due to laptop use at flexible workplaces inside an office: case-report of a bilateral ulnar neuropathy

(continued)

Table 12.1 (continued)

Author	Title
Larrea-Araujo, C., Ayala-Granja, J., Vinueza-Cabezas, A., Acosta-Vargas, P. (2021)	Ergonomic risk factors of teleworking in ecuador during the COVID-19 pandemic: a cross-sectional study
Leroy, S; Schmidt, AM and Madjar, N (2021)	Working from home during COVID-19: a study of the interruption landscape
Matsugaki, R.; Muramatsu, K.; Tateishi, S[J]; Nagata, T[J]; Tsuji, M; Hymn, A; Ikegami, K; Fujino, Y[J]; Matsuda, S (2021)	Association between telecommuting environment and low back pain among Japanese telecommuting workers a cross-sectional study
Palumbo, R (2020)	Let me go to the office! An investigation into the side effects of working from home on work–life balance
Palumbo, R; Manna, R; Cavallone, M (2021)	Beware of side effects on quality! Investigating the implications of home working on work–life balance in educational services
Radulović, A.H., Žaja, R., Milošević, M., (…), Luketić, I., Božić, T. (2021)	Rad od kuće i mišićno-koštani bol tijekom epidemije COVID-19 - Pilot-istraživanje I [work from home and musculoskeletal pain in telecommunications workers during COVID-19 pandemic: a pilot study]
Raisiene, AG; Rapuano, V; Varkuleviciute, K; Stachova, K (2020)	Working from home-who is happy? A survey of Lithuania's employees during the COVID-19 quarantine period
Ralph, P., Baltes, S., Adisaputri, G., (…), Milani Fard, A., Alkadhi, R. (2020)	Pandemic programming: how COVID-19 affects software developers and how their organizations can help
Restrepo, BJ; Zeballos, E (2020)	The effect of working from home on major time allocations with a focus on food-related activities
Šagát, P., Bartík, P., González, P.P., Tohănean, D.I., Knjaz, D. (2020)	Impact of COVID-19 quarantine on low back pain intensity, prevalence, and associated risk factors among adult citizens residing in Riyadh (Saudi Arabia): a cross-sectional study
Sanchez, DG; Parra, NG; Ozden, C; Rijkers, B; Viollaz, M; Winkler, H (2020)	Who on earth can work from home?
Schall, M.C., Chen, P. (2021)	Evidence-based strategies for improving occupational safety and health among teleworkers during and after the coronavirus pandemic
Seva, R.R., Tejero, L.M.S., Fadrilan-Camacho, V.F.F. (2021)	Barriers and facilitators of productivity while working from home during pandemic
Sousa-Uva, M., Sousa-Uva, A., and Sampayo, M.M., Serranheira, F. (2021)	Telework during the COVID-19 epidemic in Portugal and determinants of job satisfaction: a cross-sectional study sectional study
Tavares, F., Santos, E., Diogo, A. Ratten, V. (2020)	Teleworking in Portuguese communities during the COVID-19 pandemic

(continued)

Table 12.1 (continued)

Author	Title
Toniolo-Barrios, M; Pitt, L (2021)	Mindfulness and the challenges of working from home in times of crisis
Zhang, SH; Moeckel, R; Moreno, T.A., Moreno. Shuai, B; Gao, J (2020)	A work–life conflict perspective on teleworks
Zyznawska, J.M., Bartecka, W.M. (2021)	Remote working forced by COVID-19 pandemic and its influence on neck pain and low back pain among teachers I [wpływ pracy zdalnej wymuszonej pandemią COVID-19 na ból kręgosłupa w grupie zawodowej nauczycieli]

Source Own elaboration

According to BELZUNEGUI-ERASO and Erro-Garcés (2020), one of the effects of this in the world was the adoption of telework. So, the adoption of telework has grown as a rapid option to reduce exposure to the SARS-COV virus in the pandemic should not be extinguished at the end of it, but rather exist within the new normal with the support of technologies such as high-speed internet, cloud file storage, new technologies, and communication software ILO (2020).

A dilemma arose with the shutdown of economic activities: how to protect people's health without damaging the economy? For this, the governments of several countries imposed on their entrepreneurs, when possible, the adoption of full-time telework, this requirement left truly little time for organizations to adapt to the new model (ILO 2020).

Telework is presented as work that is performed anywhere that is not usual. And home office, it is a telework and means that it is done partially or totally in the worker's house (ILO 2020).

Pre-pandemic studies point out that remote work brings with it the concept of flexibility, which gives the employee a better quality of life by allowing better management of their time devoted to work and personal life, and this results in improved productivity (Bloom et al. 2016).

The company needs to incorporate ergonomic concepts to ensure the long-term health of workers, especially when it comes to telecommuting (Davis et al. 2020).

However, Goldenberg et al. (2022) present the most common telework-related challenges as being: dissatisfaction with technology/software, work arrangements, ergonomics, work–life balance, communication within the organization, and uncertainties regarding career development.

Regarding the personal and/or family challenges of telework, the most cited are social isolation, the impact of the pandemic on mental health, school closings and homeschooling, care for vulnerable family members, and concerns about daycare (Goldenberg et al. 2022).

These effects on the population are generated by the relationship that became known as WORK–LIFE balance and has become more noticeable over time. It is because how the relationship between personal life and work is seen has changed from the 1970s when there was a focus on the double journey, especially of women who after performing their paid work activities still had to face intrafamily (domestic) work without remuneration, for a more plural discussion. Although more inclusive, this relationship still brought criticism because it could suggest an artificial separation between personal life and work (Pleck and Kanter 1976; Bloom et al. 2016).

Souza-Uva et al. (2021) point out that organizational culture and the work environment play a key role in teleworking satisfaction.

A work environment is not only composed of just a computer, but several other details of the workplace will also directly impact on the ergonomics of the employee. That is why employers must consider the employee, looking for both the opportunities generated by telecommuting, but also the problems involved (family situations, skills, equipment, etc.) (ILO 2020).

The negative effects of telework are those related to physical and mental health (mental distress, such as stress, anxiety, depression, and work-related physical conditions such as musculoskeletal disorders). These can occur due to isolation, limited mobility, working conditions, the number of hours worked, the number of works breaks or the lack thereof, and the difference between what the work demands and the skills of the workers (Crawford et al. 2011; Moretti et al. 2020; Vander et al. 2017).

A survey made with companies of the telecommunications sector showed that in telework, two-thirds of the employees who respond the survey have said that they have suffered more musculoskeletal problems in this type of work compared to face-to-face work, such as stronger pain in the lumbar region, upper back/neck, and hands, with a higher incidence in women (Radulović et al. 2021).

Some employees mention they do not have their own space for work at home, some say they do not have an office desk to work on, a small part says they use an ergonomic chair, and more than half of the survey participants say they work more hours at home than they do at the company (Radulović et al. 2021).

Some strategies used to combat the stress caused by telecommuting are exercise, spending time outdoors, spending time with family members or friends. To improve some other situations suffered by the teleworker, the recommendations for organizations are based on the improvement of their employees' technological capabilities, have objective and clear communication, and provide the necessary hardware and software to ensure comfortable ergonomics, flexibility in schedules, decisions regarding return to work, and increasing benefits related to childcare (Goldenberg et al. 2022).

Cuerdo-Vilches et al. (2021) state that the exclusive and fixed-set spaces set to work were presented as the most appropriate, as well as the best-defined digital resources. It also mentioned the larger spaces, with comfortable temperatures, natural or artificial lighting, and windows, the places with better soundproofing, external views, more suitable furniture, among others.

But the flexibility brought by telework brings burden and bonus. If on one hand, the employee eliminates a real waste of time such as what was used to be spent in commuting, they start to use as a work environment a place that was often not planned for this, with inadequate levels of noise, lighting, and inadequate for the human body with issues such as the height of chairs and tables. In addition, there are issues analyzed by ergonomics such as the mental workload and the fact that the employee is always available to meet the demands of the organization, which causes a very great physical exhaustion (Davis et al. 2020).

As the home office becomes a permanent option for many workers, it needs to incorporate ergonomic concepts to ensure the long-term health of workers, so it is incumbent on corporations to provide employees with proper ergonomic equipment and information, or else what is currently uncomfortable will evolve into diseases and muscle-skeletal disorders, which will directly affect the bottom line with treatment costs, absenteeism, and other compensation costs (Davis et al. 2020).

12.5 Case Study

Case 4.1: Novo Banco—Sustainability Report (2020)

With the pandemic, in 2020 there was an increase in the challenge of "Caring for the health, well-being, and safety at work" of Novo banco's human capital.

At first, the resources focused on emergency response, where the priority was to protect people and implement actions that conveyed a sense of security. With this, many employees of the company were placed for Telework without jeopardizing the organization's activity and safety.

In this scenario, the bank puts a significant part of its employees working in an office.

In the commercial network, about two-thirds of the employees were present (commercial functions do not allow their full performance on a telework regime), and in the central departments, this value was approximately 25%. Because of this, the employees were within any of the three work contexts below:

- Always in office.
- Always in telework (primacy for risk groups and pregnant women).
- In rotation.

For employees who performed their duties in the telework regime, NOVO BANCO has prepared a set of tips and best practices for better adaptation to new forms of work, which covers not only good individual practices, but also team management and safety, to guarantee that everyone has a physical and mentally safe and healthy work environment.

In the scope of occupational safety, and the specificity of the pandemic situation, in 2020 the NOVO BANCO provided audits of its central buildings where many people gather to the cafeterias to some branches, with the focus on assessing the implementation of procedures and practices in the context of the pandemic. Concomitantly, the bank followed up on the assessment of risks related to working conditions and the functions performed.

In 2020, a survey was carried out where about 2.4 thousand people answered questionnaires about satisfaction with attendance and the results show that 89% of medium-sized company customers and 87% of large-company customers are very satisfied with the bank's attendance, which shows that the bank's performance is up to the needs and expectations of its customers even though many employees are working with telecommuting.

The confidence index of Novo Banco, in 2020, positioned at 75% and the overall satisfaction maintained a positive evolution, where the weight of very satisfied customers, in the same year, placed at 75% compared to 69% achieved in 2019.

In June and July, analyses were made to the companies to evaluate the measures and solutions adopted by NOVO BANCO in the scope of COVID-19, because of the clients' needs and expectations, and 93% of the clients the company indicated that these measures were in accordance with or above expectations. It is observed that 71% of the clients were very satisfied with the support granted by NOVO BANCO in this period.

Case 4.2: The Source of This Case Study—ILO (2020)

The Malta Government's Economic Development Agency (Malta Enterprise—ME) is responsible for attracting foreign investment to Malta and ensuring that companies operating from Malta have a business environment suitable for innovation, consolidation, and expansion of their economic activity.

The master policy adopted by the Maltese government as a response to the COVID-19 pandemic was remote working, which together with other flexible working arrangements was already gaining ground before the crisis.

Key measures to contain the spread of the COVID-19 virus are encouraging people to work from home, limiting interactions between individuals from different families, and preventing meetings in the workplace.

The Maltese government through Malta Enterprise noted that the promotion and adoption of telecommuting during this period could positively change the way organizations view telecommuting. Furthermore, the increasing acceptance of telework during the COVID-19 pandemic tends to lead to higher rates of female labor force participation in future.

Data from the Maltese National Statistical Office, through its Labor Force Survey, shows that in 2019, on average, 12 percent of the employed population was teleworking; this percentage had risen to 33 percent by the end of March 2020. A further increase is expected by April 2020.

How to Facilitate Teleworking Activities

On March 16, 2020, Malta Enterprise (ME) issued a guidance document and an application form under the Business Development and Continuity Program; such document was made with a bias to facilitate employers to make telecommuting modalities available so that their workforce could perform their work from home. The competition arising from this program supported both organizations and self-employed workers in investing in technologies that would allow teleworking, partially covering the costs of teleworking solutions.

The program was defined as: support limited to five hundred euros per teleworking agreement and 4,000 euros per company. Overall, the incentives granted covered 3,740 teleworking arrangements. Malta Enterprise estimates that the amount awarded to promoters under this program totals about €1.2 million.

Main Challenges and Lessons Learned

Malta Enterprise employs about one hundred people and acts as a point of contact for companies regarding all the incentives that the Government has launched about COVID-19.

For ME, this task was not always easy, as the volume of applications for review was Teleworking during and after the COVID-19 pandemic.

Practical Guide

The time to respond to any questions and the provision of assistance were critical. The main challenge was the effective allocation of resources to all the programs implemented.

For now, wherever possible, we continue to encourage people to continue working from their homes. This measure is crucial to ensure that the risk of contagion is kept to a minimum. The example from the Government comes from the authorization for the public sector to continue to work from home that has been extended. This approach for private companies operating in sectors where teleworking is allowed is being followed. A consensus has been created in Malta that, wherever possible, people should remain in telework.

Case 4.3: Coca-Cola Brazil—Sustainability Report (2020)

In March 2020, all Coca-Cola Brasil employees began working from home. With social isolation, the remote work, initially planned for three months, lasted through 2020 and will continue in 2021 as a safety measure and way to minimize the spread of the disease by avoiding the displacement of professionals whose responsibilities do not require physical presence within the company.

All employees received information about safety and hygiene measures, as well as guidance on healthy habits, mental health, and practices to improve the home office experience and routine. The health and well-being guidelines were shared with the entire Coca-Cola Brazil System.

The company's main executives and employees meet virtually on weekly basis, in online meetings that ensure everyone is incorporated and updated about the changes and priorities of this new circumstance. As time passed and the pandemic extended, information was placed on a single channel and shared by Yammer, the company's social network. A Coca-Cola Brazil newsletter was also produced in a Friday edition—the newsletter has become a relevant communication channel in the company.

When measuring the degree of customer satisfaction, the Coca-Cola Brazil System also achieved a score considered high in all communication channels in 2020, which positioned the organization in the excellence zone of this measure.

Case 4.4:.The Home Office: Ergonomic Lessons from the "New Normal—Davis et al. (2020)"

An ergonomic assessment was conducted via a survey sent to all professors (N = 4200) and staff (N = 4300) at the University of Cincinnati, which included an email address to submit photos upon completion of the survey so that participants could have their home office assessed by an experienced ergonomist.

The instructions requested two photos of the individual working, one directly behind the person using the computer in the home office area and one from the side. One individual (KGD) provided the initial ergonomic assessment, identifying concerns and recommendations. Another individual (SEK) reviewed and edited the assessments, providing any entries where they felt something was missing. The other team members coded and rated the features of each home office, which were then reviewed by KGD. The following components were rated: (1) monitor (primary/ secondary, laptop/external, very high/very low, centralized/decentralized), (2) chair (chair type, hard seat, very high/very low, five casters), armrests, armrests worn, armrests adjusted correctly, lumbar support, feet on the floor), (3) work surface (hard front edge, bright, very dark, task light), (4) input devices (external keyboard, laptop keyboard, laptop touchpad) or mouse, external mouse), and (5) workstation type (sitting, standing, other). The data was then summarized across all workstations. The recommendations focused on feasible and inexpensive fixes for most of the identified concerns. All recommendations were then reported to the individuals, along with encouragement to implement them.

The ergonomic assessments of the workstations identified many problems that could be negatively affecting the workers, including chairs being the wrong height (41% too low and 2% too high), which resulted in raised arms, front edge tilt, and poor head position. A common characteristic of the chairs was a solid surface to sit on (63%). Three of the workers were sitting at a non-office workstation, such as a sofa or bed for their main work area. A small majority of workers had armrests on their chairs (53%), but many people did not use them (32%), or the armrest was poorly adjusted (18%). The lack of use of armrests causes contact stress on the forearms and strain on the upper back. The chair backrest support was often not used (69%) and often without lumbar support (73%).

Thus, many individuals do not have adequate lumbar support (Table 12.2).

12.6 Indication of How the Conceptual Framework Will be Incorporated into the ESG development, Deployment, and Monitoring Process of ESG

According to the literature, it is understood that some suggestions of actions regarding health and quality of life for the worker can be adopted by companies. Below are suggested actions:

Table 12.2 Text document—summary of features for chair, monitor, and workstation

Chair height		*Monitor height*	
Chair too high	1	Primary external monitor too high	1
Chair too low	17	Primary external monitor correct height	10
Chair at right height	19	Primary external monitor too low	12
N/A	4	Secondary external monitor too high	1
Seat of chair		Secondary external monitor correct height	1
Hard seat	25	Secondary external monitor too low	5
Seat with cushion	12	Laptop monitor too low	30
Couch/bed	3	*Monitor location*	
Armrest		Primary external monitor centered	12
Armrests present	22	Primary external monitor off centered	8
No armrests	15	Secondary external monitor centered	4
Armrests used	7	Secondary external monitor off centered	11
Armrests properly adjusted	4	Laptop monitor centered	21
Back support of chair		Laptop monitor off centered	7
Lumbar support	8	*Other considerations*	
No lumbar support	30	Feet not on floor	3
Back against chair	11	Task light	25
Back not against chair	25	Glare	7
Edge of workstation surface		Too dark	3
Hard front edge	37		
Rounded edge	1		

Source Davis et al. (2020)

- The company cannot control the design and ergonomic properties inside the employee's home due to legal issues related to private property and privacy. However, it is possible to offer to supervise the place where the work is performed at home and to provide an ergonomic desk, ergonomic chair, and knowledge on how to improve working conditions and make them ergonomically acceptable (Radulović et al. 2021).
- Companies should promote frequent breaks and exercises (Radulović et al. 2021).
- Make sure that employees work within their workday, without exceeding the time they would do in the office (Radulović et al. 2021).
- The employer supplies support and services for career advancement (Goldenberg et al. 2022).

- More understanding managers with greater empathy for employees' needs, allowing more flexibility to meet the needs of every individual within what is possible. This includes identifying reduced work capacity and accommodating individual needs and circumstances (Goldenberg et al. 2022).
- Managers more understanding and with greater empathy for the needs of employees, allowing more flexibility to meet the needs of every individual within what is possible. This includes identification of reduced working capacity and accommodation of individual needs and circumstances (Goldenberg et al. 2022).
- More exclusive spaces and those spaces set fixedly to work were presented as the most suitable, as were better defined digital resources, larger spaces with comfortable temperatures, natural or artificial lighting, and windows, the places with better sound insulation, external views, and more suitable furniture, among others (Cuerdo-Vilches et al. 2021).
- The responsibilities and rights of each party (employee and employer) must be established (Cuerdo-Vilches et al. 2021).
- Depending on the sector affected, they need to have the support of the government, and other social collectives for the provision of resources or promotion of alternative measures to ensure this work model, as well as the reconciliation of family and work in the most feasible way (Cuerdo-Vilches et al. 2021).
- When providing computer equipment, characteristics of the employee and the work environment he/she has at home (Davis et al. 2020) must be considered.

Ensure the level of skills for access to and use of these technologies (ILO 2020).

- Review the policy regarding permission access to data and work servers from personal equipment or the possibility of receiving ICT equipment provided by the employer (ILO 2020).
- The inappropriate use of laptops brings severe ergonomic limitations such **as** inadequate screen and keyboard height (Davis et al. 2020).
- Use of soft and cylinder-shaped materials on the edge of the tables: to avoid injury by contact with rigid and straight surfaces (Davis et al. 2020).
- The use of individual telework plan, which should be discussed and agreed upon with direct supervision (ILO 2020).
- Agree on a common system to signal availability for work and ensure that both management and colleagues respect it (ILO 2020).
- Encourage employees to communicate whenever they feel overloaded, act as an early warning to detect the risk of burnout and determine when tasks or team members need to be relocated (ILO 2020).
- Perform a mapping of personal competencies, so that the task is performed by a qualified person and adequate division of labor so as not to occur overload for any team (ILO 2020).
- Talk about how the work is done. Question people about how they work when they feel more productive energized and focused. Encourage to work in this way as much as possible (ILO 2020).

- Encourage them to share examples of changes they have made to their daily routines that have worked well (ILO 2020).
- Recognize that offline time is necessary for substantive work to be done well (ILO 2020).
- Technostress in addition to technologies and work overload with tasks increase fatigue, irritability, and the inability to disconnect from work and rest properly, so it is important to minimize them (ILO 2020).
- Eliminate the consumption of alcohol and other recreational drugs because their use reduces performance and contributes to increased aggression and violence (ILO 2020).
- Consider allowing the borrowing of office equipment for the duration of pandemic-related telework, such as office chairs, monitors, and other relevant tools that are fully compliant with ergonomic standards (ILO 2020).
- Slow or erratic internet solutions and technologies can also cause frustration and irritability; therefore, properly functioning and adequate tools should be ensured for those working remotely (ILO 2020).
- Make clear the responsibilities of employers regarding health and safety protection in telework, paying attention to health and safety hazards and risks, homework environment conditions, equipment, ergonomic factors, stress, as well as other mental health issues (ILO 2020).
- Make clear the rights and responsibilities of teleworkers regarding their health and safety when they work from home (ILO 2020).
- Update and equip occupational safety and health (OSH) professionals with tools and processes regarding worker health support systems, work environment, training, information, and compliance monitoring mechanisms regarding OSH and ergonomic protocols specifically designed for those in telework (ILO 2020).
- Ensure the existence of easy access to psychological support options for employees to confidentially share their fears and concerns, i.e., through access to helplines, counseling, assistance programs (ILO 2020).
- Encourage management to set an example for people under their supervision with actions that show ways to mitigate stress (ILO 2020).

12.7 Summary of Main Ideas and Concepts

- Teleworkis the flexible way of providing services that occur due to the technological revolution experienced in recent years and have led to the development of virtual relationships, allowing work to be performed anywhere other than the company's environment (ILO 2020).
- Home office—a telework that is done partially or totally at the worker's home (ILO 2020).
- Ergonomics—is the scientific discipline that deals with understanding the interactions between humans and other elements of a system and uses theory, principles, data, and methods as planning to optimize human well-being and the performance

of that entire system. The terms ergonomics and human factors are often used and consider physical, cognitive, sociotechnical, organizational, environmental, and other pertinent factors and the complex interactions between humans, the environment, tools, products, equipment, and technology (IEA n/d).

12.8 Questions for Reflection and Study

1. What are the advantages for the company to adopt ergonomic care for the teleworker?
2. What are the disadvantages for the company to adopt ergonomic care for the teleworker?
3. Why, despite seeming more advantageous, do many companies still not adopt ergonomic care?
4. What actions can be used to monitor the ergonomics performed by the company?
5. What ergonomic measures can we cite as important for the teleworker's quality of life?
6. Whose responsibility is ergonomics in telecommuting?
7. How can an organization teach the concepts of ergonomics to employees remotely so that they can apply them in their workspace?

12.9 Cases to be Resolved

1. An institution that adopts telecommuting and most of its employees have small children, what measures can the company take so that the employee can feel safety, tranquility, time, and satisfaction to produce in the best possible way?

How can a company that uses a home office contribute to a correct adaptation of the physical space for the employee inside his residence?

References

Abellán García A, García MA, Legaz-García MC (2021) The impact of corporate social responsibility on financial performance: evidence from the Spanish food industry. Sustainability 13(2):659

Alshehhi A, Nobanee H, Shahwani HA (2018) The impact of sustainability practices on corporate financial performance: literature trends and future research potential. Sustainability 10(6):1989

Bloom N, Liang J, Roberts J, Ying ZJ (2016) Does working from home work? Evidence from a Chinese experiment. The Quarterly Journal of Economics, 130(1):165–218

Carlos V, Morais AI (2017) Corporate social responsibility communication in the banking sector: a literature review. Corporate Communications: An International Journal 22(4):519–537

Crawford ER, LePine JA, Rich BL (2011) Linking job demands and resources to employee engagement and burnout: a theoretical extension and meta-analytic test. J Appl Psychol 95(5):834
Cuerdo-Vilches T, Navajas-Romero V, del-Castillo-Feito C (2020) Responsible communication in social media: a review of the literature on ethical and transparent approaches. Communication & Society 33(4):117–130
Cuerdo-Vilches T, Navajas-Romero V, del-Castillo-Feito C (2021) Corporate social responsibility communication in social media: a proposal for a model of ethical and transparent practices. Sustainability 13(10):5489
Davis K, Farrell L, Stokes L (2020) Does working from home work? Evidence from returns to homeownership. Reg Sci Urban Econ 80:103487
Elkington J (1994) Towards the sustainable corporation: win-win-win business strategies for sustainable development. Calif Manag Rev 36(2):90–100
Elkington J (2018) 25 years ago i coined the phrase triple bottom line. Here's why it's time to rethink it. Harvard business review. Available at: https://hbr.org/2018/06/25-years-ago-i-coined-the-phrase-triple-bottom-line-heres-why-im-giving-up-on-it. Accessed 25 Nov 2020
Erro-Garcés A (2020) Communication for development and social change: a critical review of the literature. Communication & Society 33(1):1–14
Goldenberg MA, Mainiero LA, Sullivan SE, Forret ML (2022) Exploring the bright and dark sides of remote work: a review and research agenda. J Manag 48(1):18–50
Hutchins MJ, Sutherland JW, Dornfeld DA (2019) An exploration of manufacturing paradigms for sustainable development. J Manuf Syst 50:137–148
ILO (2020) World employment and social outlook: trends 2020. International Labour Organization
Moretti E, Giorcelli F, Guiso L (2020) The productivity effects of working from home: evidence from a panel of Italian firms. National Bureau of Economic Research Working Paper, (27917)
OPP (2017) Manual de Oslo 2018: Directrices para recoger, informar y usar datos de innovación. OECD Publishing
Oželienė V (2018) The concept of sustainable development and its impact on the development of corporate social responsibility. Entrep Sustain Iss 6(2):643
Pleck JH, Kanter RM (1976) Work-family conflict. J Soc Issues 32(3):163–178
Radulović B, Kovačević V, Ljumović I (2021) The impact of work from home on worklife balance: a systematic literature review. Sustainability 13(16):9239
Savitz AW (2007) A empresa sustentável: o verdadeiro sucesso é lucro com responsabilidade social e ambiental. Afonso Celso da Cunha Serra (trad.). Elsevier, Rio de Janeiro
Soschinski A, Fröhling M, Thiede S (2019) Towards a holistic approach for sustainable manufacturing: a systematic literature review. Procedia CIRP, 80:25–30
Souza-Uva V, Pimenta ML, Silva SL (2021) Work-life balance and teleworking: a systematic literature review. Sustainability 13(11):6061
Vander Elst T, Verhoeven C, Van den Broeck A, Baillien E (2017) Not working from home but always being at work: how boundaryless work blurs the line between work and private life. Neuropsychiatric disease and treatment 13:2821

Correction to: ESG Challenges in the Social Context of the Animal Protein-Based Food Sector: The Brazilian Case

Juliana Damaris Candido De Lima, Renan Carriço Payer, Samuel Martins Drei, Lizzie Bessa Risicato, Ruben Huamanchumo Gutierrez, Lidia Angulo-Meza, Gilson Brito Alves Lima, Lincoln Campelo Dias, and Pâmela De Carvalho Marques Silva

Correction to: Chapter 7 in: M. J. Meiriño and O. L. G. Quelhas (eds.), *Critical Factors Associated with Environmental, Social and Governance Management in Organizations*, Accounting, Finance, Sustainability, Governance & Fraud: Theory and Application, https://doi.org/10.1007/978-981-96-1725-8_7

In the original version of the book, the following belated corrections have been incorporated: The author name "Renan Cariço Payer" has been changed to "Renan Carriço Payer" in the Frontmatter, Backmatter and in Chapter 7. The book has been updated with the changes.

The updated version of this chapter can be found at https://doi.org/10.1007/978-981-96-1725-8_7

M. J. Meiriño and O. L. G. Quelhas (eds.), *Critical Factors Associated with Environmental, Social and Governance Management in Organizations*, Accounting, Finance, Sustainability, Governance & Fraud: Theory and Application, https://doi.org/10.1007/978-981-96-1725-8_13

Correction to: ESG Challenges in the Social Context of the Animal Protein Based Food Sector: The Brazilian Case

[illegible]

[illegible]

Author Index

M. J. Meiriño and O. L. G. Quelhas (eds.), *Critical Factors Associated with Environmental, Social and Governance Management in Organizations*, Accounting, Finance, Sustainability, Governance & Fraud: Theory and Application, https://doi.org/10.1007/978-981-96-1725-8

Subject Index

M. J. Meiriño and O. L. G. Quelhas (eds.), *Critical Factors Associated with Environmental, Social and Governance Management in Organizations*, Accounting, Finance, Sustainability, Governance & Fraud: Theory and Application, https://doi.org/10.1007/978-981-96-1725-8

The manufacturer's authorised representative in the EU is Springer Nature Customer Service Centre GmbH, Europaplatz 3, 69115 Heidelberg, Germany. If you have any concerns regarding our products, please contact ProductSafety@springernature.com

Printed and bound by CPI Group (UK) Ltd, Croydon, CR0 4YY
01/10/2026
02225928-0003